Statistical Process Control in Automated Manufacturing

QUALITY AND RELIABILITY

A Series Edited by

Edward G. Schilling

Center for Quality and Applied Statistics
Rochester Institute of Technology
Rochester, New York

1. Designing for Minimal Maintenance Expense: The Practical Application of Reliability and Maintainability, *Marvin A. Moss*
2. Quality Control for Profit, Second Edition, Revised and Expanded, *Ronald H. Lester, Norbert L. Enrick, and Harry E. Mottley, Jr.*
3. QCPAC: Statistical Quality Control on the IBM PC, *Steven M. Zimmerman and Leo M. Conrad*
4. Quality by Experimental Design, *Thomas B. Barker*
5. Applications of Quality Control in the Service Industry, *A. C. Rosander*
6. Integrated Product Testing and Evaluating: A Systems Approach to Improve Reliability and Quality, Revised Edition, *Harold L. Gilmore and Herbert C. Schwartz*
7. Quality Management Handbook, *edited by Loren Walsh, Ralph Wurster, and Raymond J. Kimber*
8. Statistical Process Control: A Guide for Implementation, *Roger W. Berger and Thomas Hart*
9. Quality Circles: Selected Readings, *edited by Roger W. Berger and David L. Shores*
10. Quality and Productivity for Bankers and Financial Managers, *William J. Latzko*
11. Poor-Quality Cost, *H. James Harrington*
12. Human Resources Management, *edited by Jill P. Kern, John J. Riley, and Louis N. Jones*
13. The Good and the Bad News About Quality, *Edward M. Schrock and Henry L. Lefevre*
14. Engineering Design for Producibility and Reliability, *John W. Priest*
15. Statistical Process Control in Automated Manufacturing, *J. Bert Keats and Norma Faris Hubele*

Additional volumes in preparation

Statistical Process Control in Automated Manufacturing

edited by

J. Bert Keats
Norma Faris Hubele
Arizona State University
Tempe, Arizona

Marcel Dekker, Inc. New York and Basel

Library of Congress Cataloging-in-Publication Data

Statistical process control in automated manufacturing / edited by J. Bert Keats, Norma Faris Hubele.
p. cm.—(Quality and reliability; 15)
Includes index
ISBN 0-8247-7889-8
1. Quality control—Statistical methods. 2. Process control—Statistical methods. I. Keats, J. Bert (John Bert) II. Hubele, Norma Faris. III. Series
TS156.S757 1989
670.42–dc19 88-13113

MARCEL DEKKER, INC.
270 Madison Avenue, New York, New York 10016

Current printing (last digit):
10 9 8 7 6 5 4 3 2 1

PRINTED IN THE UNITED STATES OF AMERICA

About the Series

The genesis of modern methods of quality and reliability will be found in a simple memo dated May 16, 1924, in which Walter A. Shewhart proposed the control chart for the analysis of inspection data. This led to a broadening of the concept of inspection from emphasis on detection and correction of defective material to control of quality through analysis and prevention of quality problems. Subsequent concern for product performance in the hands of the user stimulated development of the systems and techniques of reliability. Emphasis on the consumer as the ultimate judge of quality serves as the catalyst to bring about the integration of the methodology of quality with that of reliability. Thus, the innovations that came out of the control chart spawned a philosophy of control of quality and reliability that has come to include not only the methodology of the statistical sciences and engineering, but also the use of appropriate management methods together with various motivational procedures in a concerted effort dedicated to quality improvement.

This series is intended to provide a vehicle to foster interaction of the elements of the modern approach to quality, including statistical applications, quality and reliability engineering, management, and motivational aspects. It is a forum in which the subject matter of these various areas can be brought together to allow for effective integration of appropriate techniques. This will promote the true benefit of each, which can be achieved only through their interaction. In this sense, the whole of quality and reliability is greater than the sum of its parts, as each element augments the others.

The contributors to this series have been encouraged to discuss fundamental concepts as well as methodology, technology, and procedures at the leading edge of the discipline. Thus, new concepts are placed in proper perspective in these evolving disciplines. The series is intended for those in manufacturing, engineering, and marketing and management, as well as the consuming public, all of whom have an interest and stake in the improvement and maintenance of quality and reliability in the products and services that are the lifeblood of the economic system.

The modern approach to quality and reliability concerns excellence: excellence when the product is designed, excellence when the product is made, excellence as the product is used, and excellence throughout its lifetime. But excellence does not result without effort, and products and services of superior quality and reliability require an appropriate combination of statistical, engineering, management, and motivational effort. This effort can be directed for maximum benefit only in light of timely knowledge of approaches and methods that have been developed and are available in these areas of expertise. Within the volumes of this series, the reader will find the means to create, control, correct, and improve quality and reliability in ways that are cost effective, that enhance productivity, and that create a motivational atmosphere that is harmonious and constructive. It is dedicated to that end and to the readers whose study of quality and reliability will lead to greater understanding of their products, their processes, their workplaces, and themselves.

Edward G. Schilling

Preface

As the technology of manufacturing moves toward more automated processes, quality engineering in general and *statistical process control* (SPC) in particular will assume newer and somewhat different roles than existed prior to the introduction of automated processes. Indeed, the quality engineering function will continue to monitor and suggest control concerning the actions of people, the performance of equipment, and the properties of materials in a way that will ensure that customer requirements are met. However, there will be substantial changes in the way this function is accomplished due to (1) an avalanche of information generated by the data acquisition processes accompanying the automated systems and (2) the critical need for rapid and sometimes instantaneous decisions about control of the process.

Automation means, among other things, more throughput and a reliance on carefully orchestrated sequences of precise, error-free operations. The implication for the quality engineering function is a demand for faster data handling so that the feedback loop between the detection of a fault and the application of corrective action will be closed in a time frame small enough either to avoid or to minimize disruption of the process.

With the changing environment of manufacturing, there is a demand for a renewed look at the methods and needs of the quality engineer. The purpose of this volume is to begin to describe the changing environment from the perspective of the quality function. Increasing demands are being placed on quality engineers to use the latest technologies to improve the competitive position of their companies. These technologies span the myriad of data-gathering and -organizing systems to complex analysis and decision-making prodecures.

This book provides a sample of the current resources and future research directions available to the quality professional. Some of the chapters are meant to provide the reader with a general description of topics, with little technical background required on the part of the reader. These will assist the quality engineer in understanding the context of quality in modern manufacturing production and decision making.

For the applied statistician interested in quality-related topics, this volume indicates the direction and activity of some of the more theoretical work. In particular, these topics address methodologies used in the modern manufacturing setting.

This text was compiled with the purpose of serving the quality profession. The editors are particularly grateful to the contributing authors who share in this purpose and vision. It is our intention and hope that the thoughts and suggestions included in this text will spur the interest of our colleagues in helping the quality engineer to keep pace with the changing, automated manufacturing environment.

J. Bert Keats
Norma Faris Hubele

Contributors

Layth C. Alwan, Ph.D. Graduate Student, Graduate School of Business, University of Chicago, Chicago, Illinois

David E. Coleman* Member of the Technical Staff, Manufacturing Technology Research Division, RCA David Sarnoff Research Center, Princeton, New Jersey

Luis E. Contreras, Ph.D. President, ProScan, Inc., Austin, Texas

Stephen V. Crowder, Ph.D. Senior Statistician, Specialty Glass and Ceramics Division, Corning Glass Works, Corning, New York

Current affiliation:

*Staff Statistical Scientist, Applied Mathematics and Computer Science Division, Alcoa Technical Center, Alcoa Center, Pennsylvania.

M. Alkan Donmez, Ph.D.* Research Associate, Department of Mechanical Engineering, The Catholic University of America, Washington, D.C.

David J. Friedman, Ph.D. AT&T Engineering Research Center, Murray Hill, New Jersey

Sakti P. Ghosh, Ph.D. Research Staff Member, Department of Computer Science, IBM Almaden Research Center, San Jose, California

K. Kumar Gidwani, Ph.D.† Manager, Department of Process Systems, LISP Machine, Inc., Los Angeles, California

Norma Faris Hubele, Ph.D. Assistant Professor, Department of Industrial and Management Systems Engineering, Arizona State University, Tempe, Arizona

J. Bert Keats, Ph.D.‡ Director, Reliability, Availability, and Serviceability Laboratory, Computer Integrated Manufacturing Systems Research Center, Arizona State University, Tempe, Arizona

Nancy J. Kirkendall, Ph.D. Senior Mathematical Statistician, Energy Information Administration, U.S. Department of Energy, Washington, D.C.

Steven R. LeClair, Ph.D. Technical Leader, Manufacturing Research, Materials Laboratory, Manufacturing Technology Division, U.S. Air Force, Wright-Patterson Air Force Base, Ohio

Douglas C. Montgomery, Ph.D. John M. Fluke Distinguished Professor of Manufacturing Engineering, Program in Industrial Engineering, University of Washington, Seattle, Washington

---

Current affiliations:

*Mechanical Engineer, Center for Manufacturing Engineering, National Bureau of Standards, Gaithersburg, Maryland.

†Senior Consultant, Department of Advanced Technology, American Express Company, Phoenix, Arizona.

‡Chairman, Department of Industrial and Management Systems Engineering, Arizona State University, Tempe, Arizona.

Jack Park* Senior Scientist, ThinkAlong Software Inc., Brownsville, California

Harry V. Roberts, Ph.D. Professor of Statistics, Graduate School of Business, University of Chicago, Chicago, Illinois

---

Current affiliation:

* Visiting Scientist, U.S. Air Force, Wright-Patterson Air Force Base, Ohio.

Contents

1
Introduction

J. Bert Keats and
Norma Faris Hubele
Arizona State University
Tempe, Arizona

The maintenance of quality has always been important in the production of goods, though methods for accomplishing this have varied over the centuries. Prior to the nineteenth century, the production of goods or manufactured items was a very labor-intensive process. Quality was assured by an apprenticeship-type structure in which production skills were judged by a close inspection of the craftsman's end product. At that time, the cost of tools was relatively low, while the number of hours required to produce a part was quite high.

In the nineteenth century, power entered the arenas of production permitting the replacement of human tasks with machines. At that time, in addition to overseeing human performance, inspection was required to ensure that the machines were correctly performing their intended function. In the late nineteenth century, the Western Electric Company and American Bell Telephone Company (presently AT&T) entered into a contractual agreement to guarantee that every instrument and material used to build components for a national communication network be thoroughly inspected (Wadsworth et al., 1986). The methods employed at that time were quite elementary and remained at that level until the 1920s, when the theory of sampling developed. Under the primary leadership of Donald A. Quarles, Walter A. Shewhart, Harold F. Dodge, and George D. Edwards, the methods and science of what was to be labeled quality assurance were established.

The control chart was introduced to the public in December 1925, in a paper by Shewhart that appeared in the *Journal of the American Statistical Association.* Its simple concept and structure were very appealing. Since workers were relied on to take the measurements on sampled output, off-line inspection was costly and resulted in time delays. However, the benefits of the improved average quality level outweighed the costs.

Shewhart's 1931 treatise presented the first unified treatment of the subject of quality control. Based on a fundamental concept of statistical control, Shewhart formulated control charts to monitor the processes within manufacturing and, when necessary, to feed back information for correcting an out-of-control situation. Using sampled data, he formulated these charts on the premise that observed process variation is attributable to either chance causes or assignable ones. Assignable causes result in unacceptable items. Their existence is signaled on a control chart as a single point outside a specified limit or a succession of points falling between some less extreme limits. With this information, those persons responsible for the manufacturing process can initiate an adjustment to the mean value, or re-center the process. In essence, the control chart was proposed as an integral part of a feedback loop used to control a manufacturing process.

Although the work of the Bell Telephone Laboratory received considerable attention in some industry and government circles (especially during World War II production), statistical methods in production were not widely applied. H. A. Freeman, an advocate of quality control and professor at Massachusetts Institute of Technology, postulated two reasons for this situation in a 1937 article:

> This slow rate of adoption probably can be explained, first, by a deep-seated conviction of American production engineers that their principal function is so to improve technical methods that no important quality variation remain, and that in any case, the laws of chance have no proper place among modern "scientific" production methods; second, by the difficulty of obtaining industrial statisticians who are adequately trained in this fairly complicated field.

Amazingly, these reasons may still apply today. Many may argue that variation in the production of goods is virtually eliminated by automation; we contend that, while the variation may be substantially reduced, assignable causes for poor quality will exist. Moreover, reduction in variability due to automation may be accompanied by more stringent tolerances as a result of competition in the marketplace. The future trend will be to remove the operator from the feedback loop and to incorporate the functions of measurement, comparison, and adjustment into the computer control.

In a 1965 review of historical developments in statistical process control Lieberman claimed that no significant progress in process control charts since Shewhart was made until A. J. Duncan's paper on the economical design of control charts was published in 1956. The Lieberman paper is strongly recommended because the adaptive control techniques it proposes for use in automated processes are still relevant today. Since the publication of the Lieberman article over 20 years ago, with the exception of efforts in the chemical processing industries, Lieberman's proposals for adaptive statistical control systems have not been heeded. Hence, the motivation for the topics in this book. Over half of the papers in this book are based on presentations made at "Statistical Process Control: Keeping Pace with Automated Manufacturing, A National Symposium," sponsored by Arizona State University and held in Tempe, Ariz., on November 6 and 7, 1986. The topic of this symposium was in response to the widening gap between automated manufacturing technology with its associated data gathering techniques and the statistics that must evaluate the data generated by these automated processes.

The data from automated processes is the result of sophisticated inspection, testing, and in-process measurement devices. Automated inspection offers the advantages of low cost, lack of human bias, rapid feedback, and an increased sampling ratio (100% in most instances). Testing differs from inspection in that it performs a functional evaluation on a device or assembly rather than ascertaining whether or not a part has been manufactured to the accuracy required in the detailed design. Automated testing is performed by providing test apparatus at all input and output points of the device being tested. The test apparatus can simulate effects such as load on the device, environmental factors, material effects, and power. Performance is measured under these conditions. Once the test apparatus is in place, desired combinations of circumstances are simulated automatically in cycles. Both steady-state and transient analyses are performed. Measuring sensors provide direct links to a computer for either immediate feedback or later study. Often, particularly with electronic systems, compensating adjustments, such as changes in resistances or capacitances, may be made to selected devices during testing to qualify a part that otherwise may not have had acceptable parameter values. In-process measurement occurs after each operation to ensure that the operation was performed correctly. If the in-process measurement reveals a problem with the immediate past operation, a second pass is made with the same part, or, if the error resulted in a defective part, another part is substituted before the operation is repeated.

The concept of quality as the "fitness for use" within an acceptable price range is filtering into the planning for production and automation. Based on the premise that data being knowledge, robots and other numerically controlled equipment are being instrumented with data collectors. These machines are intended to provide control over the manufacturing processes to increase productivity and improve quality. In summary, as the manufacturing environment changes, so will the methods for monitoring and maintaining quality.

This volume is presented in six sections:

Section I provides an introduction to statistical process control in automated manufacturing and suggests implementation strategies.
Section II focuses on time series applications in statistical process control.
Section III introduces two innovative techniques for statistical process control.
Section IV presents a paper on statistical databases for process control.
Section V explores the role of knowledge-based systems in process control.
Section VI presents a paper on how a machine tool can be controlled in real time with sensor feedback.

The 13 papers comprising this volume are individually introduced below with an accompanying commentary on the contributions of each to *statistical process control* (SPC).

1. SECTION I

Keats explains the nature of current data acquisition systems and the large volume of data which they create. The implication is that SPC systems that are capable of using "all of the data" are required in such environments. Keats mentions that although variability in automated processes will be substantially lower than corresponding variability before automation, variability will continue to be the focal point of SPC. He suggests the use of feed-forward and feedback transfer function methodology, which has enjoyed success in the chemical processing industries, for applications in manufacturing. Automated applications require a knowledge of the process in sufficient detail to develop mechanisms that will manipulate the input without human intervention when required to do so. Keats examines an area that, justifiably, has attracted little interest in SPC—the use of attributes data for control purposes. Quality engineers rely on attributes data when variables data is unavailable or too costly to obtain. With the advent of artificial intelligence, patterns of binary data may be analyzed with the assistance of expert systems to detect deviations from patterns produced under "statistical control" and to make recommendations concerning the root cause of the problem and the appropriate corrective action. The expert system may also be used to resolve conflicts when two or more statistical tests offer paradoxical results or to select the best test for the current conditions. Keats offers assistance in selecting an SPC system for automation by suggesting desirable characteristics of such systems. Although the organizations most directly associated with *computer-integrated manufacturing* (CIM) have done little to specify the role of SPC, Keats makes a few comments that should be helpful as the position of SPC is more firmly established in CIM.

Contreras provides a good introduction for the quality advocate who has the assignment of moving an organization toward implementing an on-line SPC system. He discusses issues in hardware and software, as well as difficulties of initiating organizational change. The "middle-out" implementation strategy proposed by Contreras suggests an iterative approach whereby a Master Plan is carried out by starting with a series of small incremental changes, allowing for feedback and adaptation. His expose is useful for those readers who are wondering where and how to start installing a modern SPC system.

2. SECTION II

The terminology "assignable or special causes" and "common causes" is quite familiar to the quality professional. The key to a successful control scheme is the ability to distinguish between the two in a timely fashion. Alwan and Roberts describe a study in the meaning of these terms. They suggest using the *autoregressive integrated moving average* (ARIMA) models of Box and Jenkins (1976) to capture the common cause structure of data that are not independently distributed. The residuals of these models are then examined using the traditional quality control tools to detect special causes. They illustrate their concepts by examining a well-known data set using three schemes: a traditional three-sigma control chart for independently identically distributed data, a moving-range chart, and their own proposed approach. Alwan and Roberts cite a comment made by Eisenhart in a 1963 article that appeared in the *Journal of Research of the National Bureau of Standards—C. Engineering and Instrumentation* concerning the difficulties of meeting Shewhart's conditions for "strict statistical control." With the prospect of designing an SPC system with on-line, in-process measurements, the traditional assumption of independent variables frequently appears to be violated. The analytical approach of these researchers suggests an alternative that does not depend on this assumption.

Montgomery and Friedman have conducted extensive tests that show that the Shewhart control chart for individuals is not a reliable tool when consecutive measurements are plotted. They recommend either a *cumulative sum* (CUSUM) or *geometric moving average* (GMA) chart applied to the residuals of an *autoregressive moving average* (ARMA) model. Montgomery and Friedman mention that serial data, such as data that occur in CIM, are frequently correlated, and hence they propose a model applied to the residuals. The residuals should be normally distributed and independent if the model has been correctly fitted. A multivariate approach is also described. Their results should provide impetus for the development of software that automatically fits ARMA models to data: that is, no judgments must be made by the user. Numerous reports over a wide range of disciplines have indicated that a maximum of two moving average and/or two autoregressive terms usually suffice to build a model that fits the data reasonably well. Expert systems can assist in this task. When automated ARMA building is a reality, tests for the adequacy of the model over time must be conducted. A new model must be fitted when the tests indicate inadequacy so that the decision-making process may use the best description of the current process. Such updating of models may also be accomplished automatically with the aid of knowledge-based systems.

The relationship between the Kalman filter model and the well-known exponential smoothing model is mentioned by Crowder and by Hubele in the fourth and fifth articles of this section. Jean Kirkendall provides a development and demonstration of this relationship. Based on this equivalence, the Kalman filter recursive equation is shown to be useful for implementing simple exponential smoothing and exponential smoothing with trend models. Also illustrated is the equivalence between certain ARIMA models and the Kalman filter models. As indicated by Kirkendall, an advantage of a Kalman filter approach in place of these equivalent models is the ability to start the model with no prior data. Furthermore, the model is adaptive to both known and unknown level shifts. Crowder uses this information to devise a workable SPC scheme. The exponential model with a trend is not discussed by other authors; however, it may prove useful for monitoring tool wear. Although this article does not directly address an SPC implementation, its contribution is derived from the foundation it provides linking a new SPC model, the Kalman filter, to the accepted exponential models.

The Kalman filter has seen a generation of applications in numerous areas, from classical process control problems to econometric forecasting. Crowder addresses the potential of using a form of this adaptive recursive procedure in SPC. While a considerable amount of work has been done to study the statistical aspects of the Kalman filter when the model's variance parameters are assumed known and/or constant, Crowder's estimation algorithm addresses the more practical case of having to estimate variance components and weighting factors from the data. He uses this estimation procedure, together with box and whisker plots, to formulate a control scheme for monitoring a process mean. A limited comparison using simulation of his methodology to the Shewhart approach demonstrates its superiority for detecting small shifts in the process mean. In addition to his theoretical contributions, Crowder's work is significant in that it proposes the use of an adaptive model that is useful in an SPC computer-automated environment. Model parameters may be updated in an on-line fashion and used to estimate weights for detecting an out-of-control state.

Hubele suggests two approaches for monitoring a complex system: one in the multivariate domain and the other in the stochastic. Though most of the past work in multivariate control charting has made use of the mean vector and covariance matrix components, she proposes using the individual multivariate measurements. Her work in extending narrow-limit gaging to more than one dimension illustrates the requirements of a multivariate scheme. She describes the problem in detecting out-of-control states, provides some simple rules for identifying the variable to adjust, and demonstrates the problems of using univariate control when the underlying data are multidimensional and correlated. In the second part of her paper she considers the use of a multiprocess model, which is an extension of the Kalman filter model, for process control. This model incorporates the notion of state (in-control or out-of-control) and assumes that the underlying behavior of the system may be driven by dynamics acting on the state parameters and on the observations. Though the model is somewhat complex in nature, Hubele formulates this for SPC under the assumption that it may be useful for monitoring those systems that are known to be undergoing some acceptable changes, such as the start-up of a process. Hubele's work is based on the assumption that with automation, the schemes should be selected based on the particulars of the quality environment without excessive concern for complexity of the model or rules for correction. With automation, these schemes will be performed by sensor-equipped machines, computers, and robots.

3. SECTION III

Coleman presents a novel alternative to the traditional $\bar{X}$ control charts. His proposed scheme uses the "signature" of the data by using an estimate of the cumulative distribution derived from sample data to construct a "generalized control chart." In this approach, a sequence of p-values of observed data is combined and tested against a desired threshold to indicate the incidence of unusually low or high occurrences. Coleman provides an illustrative discussion of the theoretical foundation of the method, together with several graphical ways to present the test data. His use of the glyph plot is quite useful for taking into account the time-related nature of the data. He also demonstrates the favorable performance of his scheme in comparison to the performance of some CUSUM designs. As indicated by Coleman, the significance of this work lies in its versatility of application. In addition to the example discussed in this chapter, the generalized control chart approach has potential for use with other univariate and multivariate situations. The intensive computations and sophisticated graphical displays are easily handled in a computer-intensive environment.

4. SECTION IV

Ghosh represents the perspective of the computer scientist working in manufacturing. His paper covers a broad spectrum of interesting issues in the design, storage, and retrieval of data for statistical analysis. He proposes a statistical database model that uses a relational data structure to store category attributes (e.g., a part number) and statistical attributes (e.g., a measured dimension). To reduce storage requirements, Ghosh suggests the use of statistical metadata that summarize individual observations and that may be updated with the addition of new observations. As he indicates, the requirements of a database for manufacturing and, in particular, for performing diagnoses on problems in production sometimes necessitate the storing of information on individual observations. He recommends using a sliding window to limit the amount of detailed history that is to be maintained. For those familiar with SQL/DB2, Ghosh provides some clarifying examples of how retrieval commands would be applied to his data structure. In addition to giving the reader insight into some useful computing concepts, Ghosh discusses some important properties of the statistics around which the data is organized. In closing, Ghosh also recommends analyzing the cost of data storage and retrieval as it relates to the costs and benefits of its usage in correcting manufacturing problems. This critical step in database design needs continued research.

5. SECTION V

Gidwani describes PICON (*process intelligent control*), the first expert system specifically designed for process control. The PICON system uses both hardware and software to respond in real time to large volumes of data points in a large system. The software for PICON includes both a symbolic and a numeric processor. Most of the PICON installations have been in the chemical process industries, but developments in manufacturing are already underway. The primary functions of the PICON system are those of monitoring, responding to alarm conditions, and diagnosing the root cause of the fault. The use of statistics in the PICON system is minimal. Furthermore, at the present time, the role of an expert system in SPC has yet to be defined. The following suggestions are offered for consideration: (1) the expert system functions as a pattern recognition device, comparing what is seen today with patterns from the historical database; (2) the expert system selects from among several statistical tests or procedures the one or ones that are appropriate with today's data—that is, as conditions change, so might the selected test or procedure (one test may be better under certain conditions that others); (3) when the data suggests a problem requiring corrective action, the expert system, using some weighting scheme or voting rules, combines the subjective probabilitics about the likely root causes from the domain expert (the process or quality engineer) and the past probabilities of possible sources of the problem based on relative frequencies with respect to successful fixes; and (4) the expert system determines when the parameters of the underlying model describing the data or model being used for predictive purposes should be changed. The use of knowledge-based systems in SPC is a promising and challenging opportunity. Many developments are expected throughout the next decade.

LeClair provides a taxonomy for the development of an expert system that is capable of understanding aggregate data from multiple sensors and making a qualitative evaluation of these data for control purposes. This branch of artificial intelligence, known as sensor fusion, will play an important role in multiple-input control techniques. Professionals in both artificial intelligence and process control believe that sensor fusion can produce more valid interpretations of an environment than interpretations that are made on the basis of separate analyses on each sensor. The expert system can interpret the criticality of data from each sensor, resolve conflicts when two or more sensors provide contradictory reports, and select the specific combinations of sensor responses most appropriate for the current control decision. Expert systems are suitable for sensor fusion, since, unlike their human counterparts, the sensory channels are unlikely to become overloaded and memory of past events is more dependable. Statistics have not yet been introduced as part of the sensor fusion structure; however, a few possible applications are presented here. When a sensor fails during a critical stage in a process and cannot be replaced or repaired until the stage is complete, time series techniques may be used to supply the missing observations. Similarly, such techniques may also be employed to assess the validity of values reported by each sensor. The most obvious statistical applications are those of multivariate analysis. In addition, discriminant analysis or cluster analysis techniques may be able to put sensor inputs in the appropriate categories for the current control decision.

6. SECTION VI

A cutting tool control application from research at the National Bureau of Standards is discussed by Donmez. The goal is to locate the cutting tool in the correct position with respect to the workpiece using a machine controller, a *computer numeric control* (CNC) device. Position feedback sensors report actual tool positions to the CNC controller. These signals are compared with position command signals (where the tool should be), and rigid-body kinematics are used to calculate the positional error vector of the cutting tool. Error compensation signals are then sent to a position servomechanism, which also receives information concerning lags in actual position from command position proportional to the velocity of the axis of motion. The servo then actuates movement to attain the control needed. Hundreds of error calculations are made each second. The system will not allow itself to make a bad part. This type of intrapart feedback control is quite different from feedback control after each part is completed. It is also, however, considerably more expensive. Immediate uses of this methodology may be limited to applications where precision is critical or costly scrap losses cannot be tolerated. Currently, machine tool error control is strictly deterministic. If and when statistical analysis is introduced, an obvious start would be in the analysis of the error signals.

It is hoped that this collection of papers will provide the impetus for a continued search for appropriate SPC and expert system tools for use in automated manufacturing. We feel that many of the techniques suggested in this volume are ready for use in the manufacturing arena. Others must be tested and improved upon. The next move must be made by quality professionals. Its time to take advantage of our data acquisition capabilities to improve decision-making for quality purposes.

REFERENCES

Freeman, H. A. (1937). Statistical methods for quality control. *Mechanical Engineering*, 261–262.

Lieberman, G. J. (1965). Statistical process control and the impact of automatic process control. *Technometrics*, *7*(3), 283–292.

Shewhart, W. A. (1925). The application statistics as an aid in maintaining quality of a manufactured product. *Journal of the American Statistical Association*, 546–548.

Shewhart, W. A. (1931). *Economic Control of Quality of Manufactured Product*. Van Nostrand, New York.

Wadsworth, H. M., Stephens, K. S., and Godfrey, A. B. (1986). *Modern Methods for Quality Control and Improvement*. Wiley, New York.

SECTION I

Key Issues and Implementation Strategies

2
Process Control in Automated Manufacturing: Some Key Issues

J. Bert Keats
Arizona State University
Tempe, Arizona

With sophisticated data collection systems having the capability of obtaining one or more measurements on each and every item produced virtually at the point of manufacture, *statistical process control* (SPC) techniques must adapt to keep pace. *Data acquisition systems* (DAS) are discussed. The SPC goal of determining when to adjust the process and when to leave the process alone will not change with increasing levels of automation. Likewise, patterns of variability that provide clues for separating assignable and chance causes will continue to exist with increased automation. The future of statistics in SPC is outlined with particular reference to transfer function methodologies and approaches for use with attribute data. Accuracy, speed, efficiency, adaptability, and cost-effectiveness are necessary characteristics for any SPC system which must keep pace with automated manufacturing. The role of SPC in *computer-integrated manufacturing* (CIM) is presented.

Based on a presentation made at "Statistical Process Control: Keeping Pace with Automated Manufacturing, a National Symposium," sponsored by the Center for Professional Development and the Reliability, Availability and Serviceability Laboratory, College of Engineering and Applied Sciences, Arizona State University, November 6–7, 1986.

1. INTRODUCTION—AN AVALANCHE OF DATA

In many of our factories, either operating or under construction, automated inspection and measurement systems are presenting us with the prospects associated with having information on each and every variable of interest for every item produced or process under study. We now have the capability of obtaining dozens, even hundreds, of measurements on every unit, virtually at the point of manufacture. The reality of this situation is that sophisticated data-gathering mechanisms are already in place awaiting the introduction of statistical techniques that will take advantage of having this avalanche of data.

The classical quality control charts developed prior to the advent of computers and present-day technology successfully addressed the problem of the high cost of data collection and the need for operator intervention in process adjustment. However, the potential now exists for acquiring inexpensive measurements and for implementing a system for self-adjusting machines.

With tactile systems and end-effectors, with vision systems and laser telemetry, and with voice recognition systems, we have machines that "touch," "see," and "hear" and report to computers for control purposes. The challenge is to select SPC techniques that exploit the computer-controlled environment and improve the quality of the outgoing product.

2. DATA ACQUISITION SYSTEMS

In the not-too-distant past, reference to a DAS meant a stand-alone data logger, which usually accepted analog inputs for storage on tape. A typical use of these devices would be at pumping stations for oil-field pipelines. The data, logged on tapes, would be gathered once a week and taken back to the central office for analysis. In today's automated manufacturing environment, just behind the tactile and vision sensors stands a sophisticated DAS, conditioning the data for computer analysis. In fact, the DAS may be a specialized computer itself.

There are two categories of DAS—distributed and centralized.

Carlson (1986) offers the following comparisons:

Centralized	Distributed
Connections to remote sensors are run to a central location.	Sensors are connected to front-end devices near the process.
Measured data is transferred to the computer via a parallel interface, usually the internal bus of the computer; 100,000 or more samples per second can be delivered to the computer.	Serial interfaces are used for transfer to the front end; reading rates are limited—500 or fewer samples per second.
Better suited for small, local systems; each sensor wire must go to the central DAS.	Best on large, geographically dispersed locations; a single cable is used with all sensors.
Sensor wires to the DAS may pick up stray signals.	Better measurement accuracy.

The front-end device used with distributed systems is an upgraded data logger. It may contain a multiplexer, signal-conditioning circuitry, an analog-to-digital converter, data storage circuitry, communications hardware, and the logic necessary to control each of the functions. The logger on the front end may perform control functions as well. As Roger Ladwig (1986) points out, control can range from on/off contact to PID (*proportional–integral–differential*) loop control. The process may be monitored with CRT (*cathode ray tube*) displays at the logger. Some loggers have printers to display frequency distributions, X–Y plots, and control charts. Sophisticated data loggers have menu-driven software for programming, retrieving, and displaying information. Some loggers perform array processing (including matrix algebra), fast Fourier transforms, and waveform analysis. Front-end devices are often packaged to withstand harsh environments. Digital signals are sent to a host computer. The more intelligent the front end, the less the need for a powerful host computer.

Personal computers (PCs) are increasingly popular choices for the host computer. With a PC, mathematical computations and other forms of data analysis are rather slow compared with minicomputers and mainframes. Additionally, high-speed data input must be buffered and fed to the PC at much slower rates. Hence, the extent and speed of calculations and analysis needed for process control decision-making must be considered prior to selection of the host computer.

3. VARIABILITY IN AUTOMATED PROCESSES

Traditionally, the goal of SPC has been to determine when to adjust the process and when to leave the process alone. These actions or inactions are the result of either identifying assignable causes of variability in the data or concluding that only a chance cause system of variation is present, respectively. Control charts are the primary tools for use in making these decisions.

As the technology of manufacturing moves toward more automated processes, there are a few prophets who feel that process or product variability will disappear and that the role of the quality control engineer will diminish as a result. What is more likely is that although variability will tend to decrease with advances in automation, it will not go away. What is also likely is that each decrement in variability will be accompanied by a tightening of the bounds of acceptability.

Thus, the goal of SPC will not change with technological advances in manufacturing. What will change is the specific nature of the tools used to monitor variability in the process.

4. PROCESS CONTROL GUIDELINES

Ideally, process control is proactive rather than reactive. That is, out-of-control situations can be anticipated and corrective action applied before a defective product is produced. In its purest form, a process control system is instantaneous and self-correcting. This implies that corrective action is initiated immediately after a signal indicates its need and that corrective action can be applied without human intervention.

In manufacturing today, there is a rather limited class of problems where instantaneous self-correcting control can be applied. The relationship between output and input must be well known and servomechanisms must be available. In any event, the tighter the feedback loop, that is, the shorter the response time between a signal and application of corrective action, the better the process control. More will be said about the feedback loop requirements in Section 6 of this paper.

5. SUGGESTED STATISTICS FOR PROCESS CONTROL IN AUTOMATION

Programming new automated data collection systems with traditional SPC software is failing to take advantage of all the information at hand and the potential of the new *computer-integrated manufacturing* (CIM) environment. The use of techniques that depend on sample information and the use of statistical inference should give way to methods that use all of the data available. In the very near future we should see more of the following: time series methodology (including ARIMA models and spectral analysis), regression and trend analysis, *cumulative sum* (CUSUM) techniques, geometric moving averages, multivariate approaches, sequential statistics, continuous sampling approaches, and mathematical and symbolic computer processors working in tandem for process control (expert systems with statistical databases and decision rules). A few comments will be made concerning the use of transfer functions in time series methodology. The use of sequential statistics and continuous sampling approaches will be mentioned for possible use with attributes data.

Transfer functions have been used for control purposes for over 50 years. Their use in statistical process control was introduced by Box and Jenkins (1970). A transfer function describes the relationship between the input and the output of any element in a process control loop. Until Box and Jenkins demonstrated their applicability with reference to the time series domain, their use was strictly deterministic. In the statistical treatment of transfer functions, all inputs and outputs associated with the process are expressed in terms of a time series.

Control of any process is necessary because of disturbances to the process. It may or may not be possible to measure the magnitude of these disturbances. When the major disturbances can be measured, a feedforward control model is proposed. The schematic for feedforward control is shown in Figure 1. For a specified target value for the output, the feed-forward control transfer model will indicate a change to be made in the compensating variable at time t to minimize the mean square error (sum of the squared deviations between an output value and the target value). Feed-forward control uses two transfer functions—one models the observed disturbance as a function of the output, while the other models the relationship between the compensating variable and the output. As the name implies, the compensating variable is a controllable input that recognizes the effect of disturbances in the system and attempts to compensate for their effect. Once the output has been described separately in terms of each input in a time series format, the transfer functions may be developed. The transfer functions are used to estimate the required change to the compensating variable to keep the input as close as possible to the target. When the time series model predicts an out-of-control situation t time units in the future, changes are made to the compensating variable now to offset the effects of the predicted situation.

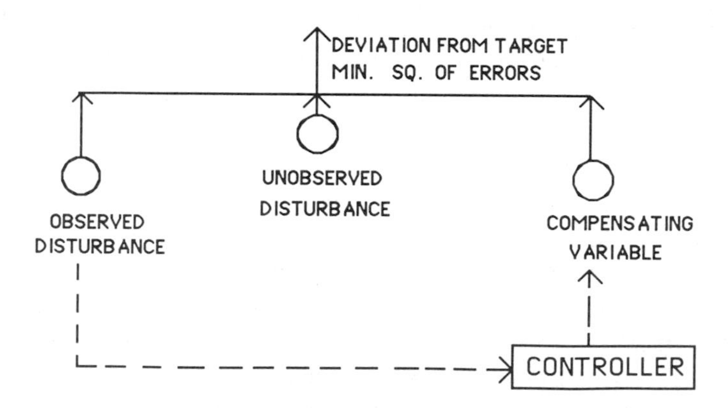

FIGURE 1 Feed-forward control model.

The nonmeasurable disturbance may be thought of as noise in the system; it indicates the presence of factors both known and unknown. The effect of system noise must be rather small to make this type of control possible. Whenever noise has a substantial effect on the output, it must be reduced either by introducing a new variable or variables that are currently part of the noise and finding ways to measure this variable or variables or by developing a feedback control schemes.

When the primary sources of disturbance are either not known or cannot be measured, feedback control may be applied. The approach uses the error signal (difference between the output and target values) as the means of identifying changes to the input. Figure 2 illustrates a feedback control situation.

Using time series analysis, the first step involves estimating the effect of all unobserved disturbances (noise) in the absence of a correcting signal. Next, a transfer function is developed linking the input (control variable) to the process and the output. This allows the control analyst to obtain the necessary adjustment to cancel the deviation from the target value due to noise.

Successful applications of transfer function models using time series analysis have been almost exclusively in the chemical process industries. Transfer functions time series models are not expected to have a substantial impact in manufacturing in the immediate future, owing to complex input/output relationships and servomechanisms that must be developed to actuate the control. However, their use in

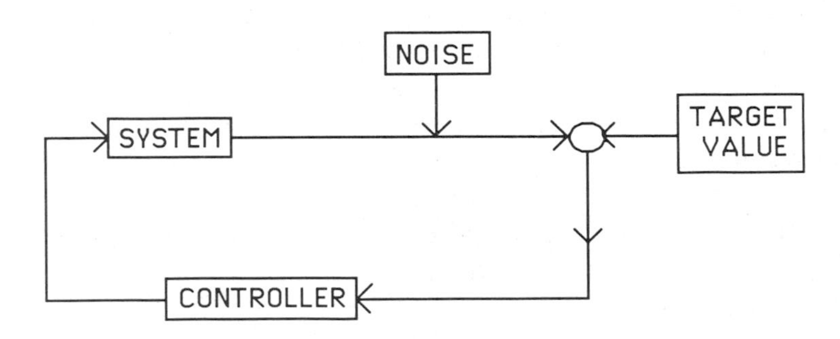

FIGURE 2 Feedback control model.

manufacturing is expected to grow steadily. The use of time series-based transfer models for control implies that statisticians and engineers trained in time series analysis and control theory are expected to be more visible SPC team members. Knowledge of the physics of the process is of critical importance in building mechanisms that establish control. Statistical experimental designs will be employed to identify the critical output variable(s) as well as the compensating (feed-forward) or control (feedback) variables. Transfer function techniques are primarily intended for real-time adaptive computer control. However, they are appropriate for use in off-line situations as well.

Very little attention has been given to process control where only attributes data are available. The traditional p, np, c, and u charts serve more of a monitoring function than one that lends itself to control since measured values are not available to provide clues for corrective action. When binary data are available (pass/fail, defective/nondefective, etc.) on each item during manufacturing, there are a few approaches that may be helpful. We can partition defects in three classes: (1) those that occur randomly and are due to chance causes of variation such as an occasional part being misfed into a machine, (2) those that appear for a while and then disappear such as defects caused by a small lot of substandard parts or defects due to an operator still learning, and (3) those that are due to a chronic problem such as a worn tool, which will not disappear until corrective action is applied. Classes (2) and (3) are said to be due to assignable causes.

Defects due to assignable causes do not occur randomly. Rather, they exhibit patterns that allow some degree of predictability. Typically, defects of Class (2) occur in clusters. Statistical analysis is capable of determining the distributions of time (measured in clock time or number of components, insertions, etc.) between clusters for different types of defects as well as identifying the distribution of defect counts within clusters. Jackson (1972) addresses this problem and presents several distributions that may be appropriate in describing time between defect clusters and defect counts within a cluster.

There are several schemes appropriate for on-line detection of a defect level change. Each was developed prior to the availability of computers for immediate analysis of data patterns for process control. Today's automated manufacturing environment is ideally suited for the introduction of one or more of these techniques. A few are mentioned below.

The *sequential probability ratio* (SPR) test developed by Wald (1947) has been primarily used in acceptance sampling; however, like many attributes acceptance sampling procedures, this test is appropriate in detecting changes in the defect level. The SPR is the ratio of the probability of a particular outcome assuming that the alternative hypothesis is valid (i.e., the true fraction defective is greater than some specified value) to the probability of the same outcome assuming that the null hypothesis is valid (i.e., the true fraction defective is equal to some specified value). The ratio is expressed in terms of the number of items examined. Items are examined or inspected one-by-one until a decision to accept or reject the null hypothesis is made. Decisions to accept (defect level remains at hypothesized value) or reject (defect level has shifted to specified higher value) the null hypothesis or to continue inspecting are based on selected probabilities of Type I or Type II errors.

Another test for a change in defect level is based on the spacing of nonconforming items. The number of conforming items intervening between the occurrence of one nonconforming item and the occurrence of the next may be represented by a geometric distribution. The higher the defect level, the smaller the spacing. The likelihood of any particular spacing value for a hypothesized defect level may be computed. If this likelihood is smaller than a specified alpha value, it may be concluded that the process has shifted to a higher defect value.

A procedure due to E. S. Page (1954), who developed the CUSUM methodology, is also appropriate. If some positive score is assigned to each defective in a group of items and a negative score is assigned to each nondefective, sums of scores may be used for use in detecting shifts in defect levels. The minimum sum of total scores for each of $n - 1$ groups collected after the last action (correction after a detected shift) was taken is subtracted from the sum of total scores for the nth group. If this value exceeds a prescribed value based on a tolerable level of fraction defective, then action should be taken.

Another possible test for a change in defect level is based on the work of Harold Dodge (1943). The expected number of items in a sequence of defective and nondefective items ending with i consecutive nondefective items is computed for some value of fraction defective that is unacceptable. If this expected number is found to be less than the actual number of items inspected in any sequence ending with i consecutive nondefective items, then it may be concluded that the defect level has increased. This is a rather crude test as the alpha level is not specified. The expected number of items in a sequence ending with i consecutive nondefective items is a convolution based on two geometric series. It is not clear whether or not the distribution of a convolution of two geometric series could be determined analytically. Computer simulation could be employed if analytic results are not easily obtainable.

Patterns of assignable causes in binary data may be identified by use of the autocorrelation statistic. Suppose that nondefective items are identified by zeros and defectives by ones. An autocorrelation of lag k value represents a correlation between observations that are k units apart. Significantly high autocorrelation values at any lag are indicative of a nonrandom pattern in the data and may be used to trigger investigations. Knowledge of the process may provide clues as to why the autocorrelation at a particular lag may be high. For example, consider a manufacturing process where four identical circuits are manufactured on the same printed circuit board. The board is subsequently sheared to yield four identical circuit boards. If order is preserved after shearing and, upon inspection, an autocorrelation at lag 4 is significant, then this is an indication that a problem exists at a particular region on the board before shearing, which may be due to the board itself or some manufacturing process that selectively affects that region of the board.

These are but a few of many possible schemes that may be applied to attributes data. Any such scheme will be of little use in process control unless it can be developed to work in tandem with someone familiar enough with the process to make inferences regarding root causes and corrective action when the attributes data suggests anomolies. Expert systems have been suggested for this purpose.

6. DESIRABLE CHARACTERISTICS OF SPC SYSTEMS FOR AUTOMATION

In selecting statistical procedures for use with automated data collection systems in manufacturing, the following characteristics must be considered:

Accuracy
Speed
Cost-effectiveness
Efficiency
Adaptability

These features are necessary for any SPC system that is to keep pace with automated manufacturing. At times, compromises may be required in the design and selection of the system due to conflicts of purpose.

Accuracy refers to how closely the model used to control the system resembles the process being controlled. The proposed model may be too complex as a result of trying to capture all of the important information for controlling the process. As a result, it may be too slow and speed may be sacrificed. The notion of data integrity whereby recorded measurements are well defined and correct is also a part of accuracy. When such measurements become prohibitively expensive, less sophisticated approaches must be adopted.

Much of the progress made in automated or computer-integrated manufacturing may be traced directly to the increases in computational speed and decreases in computing cost. With this decrease in cost has come an increase in potential of automated systems to improve productivity and quality. In quality control, this trend manifests itself in manufacturers wanting a system that is "on time" in identifying problems. No longer acceptable is a system that analyzes the sample data after the product is produced. A cost-effective system that monitors the process during production and that reports trouble spots before bad products are made is now on the agenda in many factories.

Efficiency in the utilization of information is another requirement of the preferred SPC system. With the capability of recording literally hundreds of pieces of information on every unit produced, computer storage may become cumbersome and expensive. Therefore, the SPC techniques used should be efficient in summarizing the information while still capturing the critical aspects of the process.

As we move toward a fully automated manufacturing environment, the SPC system should be designed to operate autonomously in identifying and correcting assignable causes of variation. Such a system would necessarily have to be adaptive to any changes that occur in the process.

7. RELATIONSHIPS WITH COMPUTER-INTEGRATED MANUFACTURING

The organization most associated with CIM activities is the Computer and Automated Systems Association of the Society of Manufacturing Engineers (CASA/SME). CASA/SME sponsors the annual AUTOFACT ("automated factory") Conference and Exposition as well as CIMTECH, the CIM implementation forum. An examination of literature published by CASA/SME and a review of other publications concerned with CIM has revealed very little about the role of SPC in CIM, particularly with reference to SPC in a nonsampling environment. The Technical Council of CASA/SME developed the CIM "wheel" conceptual model shown in Figure 3. The Council is composed of representatives of industry, government, and academia involved in the development and utilization of CIM technologies. Shop floor control and quality process planning are accounted for under "strategic planning" while inspection/test is a part of "factory automation." Quality engineers would, no doubt, make other interpretations about the position of quality control activities on the wheel.

Figure 4 provides a clear indication of the role of process control in the integration function of CIM. Here, process control includes any method, tool, or technique used to regulate the process "set-point" or desired value. To control engineers in a chemical or electro–mechanical environment, the term "process control" usually means PID (proportional–integral–differential) control. To the statistician or quality engineer trained in statistics, "process con-

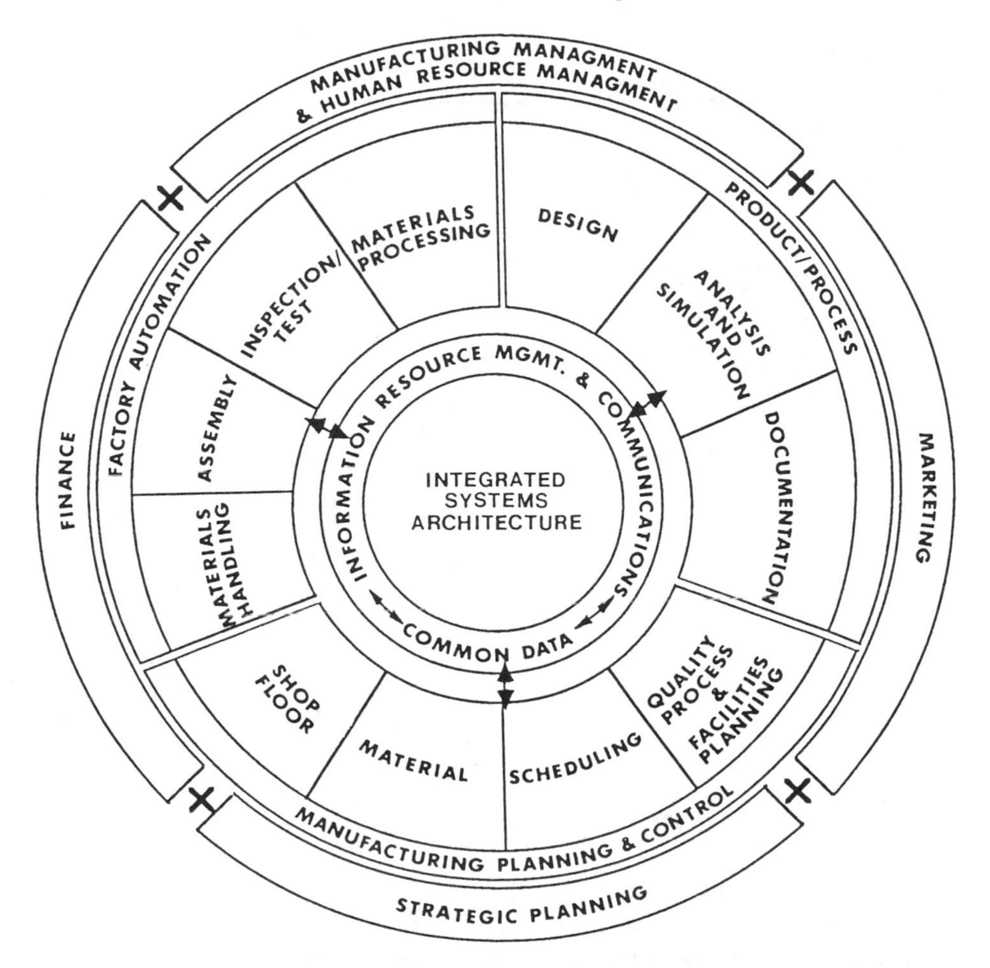

FIGURE 3 CASA/SME "CIM wheel."

trol" connotes SPC. In either case, Figure 4 provides some helpful insights. Here, SPC will be the referent of process control in the interpretation.

Information integration represents the tools necessary to create, collect, store, transport, and present information in the manufacturing system. It includes design of the product. Control integration signifies the tools directing the activities that assure that a part is being manufactured to specifications. Control is applied *prior* to

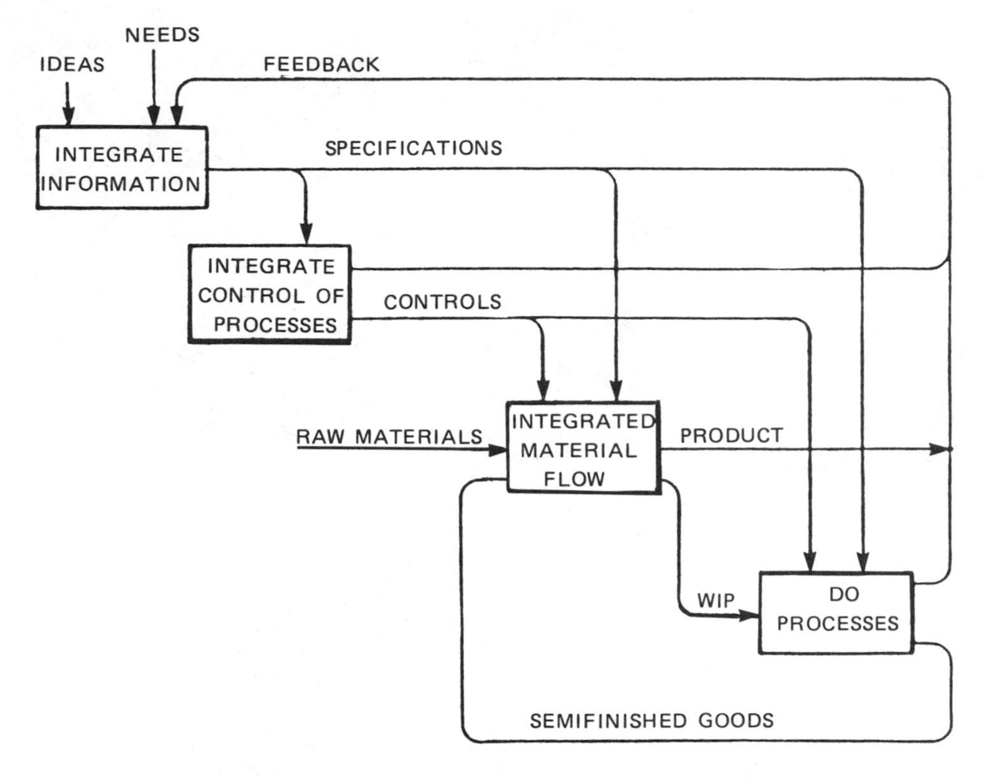

FIGURE 4 The integration role of CIM.

producing a nonconforming part. Material flow integration involves the tools necessary to grip, move, store, and place parts at the right place and at the right time.

From Figure 4, it is seen that specifications and manufacturing instructions are an input to control integration and that output from control integration provides input to information integration in the form of a signal, which describes specific needs for additional instructions. Other outputs from control integration are to material flow and processes that direct the mechanisms that are moving and performing manufacturing operations on the product at various stages.

8. SUMMARY

The present automated manufacturing environment is very different from that of the 1920s and 1930s, which gave rise to the Shewhart control charts. Instead of being scarce and costly, measurements are readily available at a relatively low cost. The challenge is to use this tremendous amount of data to improve quality. The quality professional must try nontraditional methods. Some of these methods are well developed and available. They await widespread use within the manufacturing community. Other methods must be developed. In some cases, we can take advantage of developed disciplines and creatively apply their concepts to our problems. A few of the developments have been mentioned in this paper with suggestions about implementation in an automated setting. The next move must be made by the quality community. Its time to take advantage of our data acquisition capabilities to improve decision-making for quality.

REFERENCES

Box, G. E. P., and Jenkins, G. M. (1970). *Time Series Analysis: Forcasting and Control*, Holden-Day, San Francisco (revised edition 1976).

Carlson, D. (1986). PC-based distributed data acquisition systems. *Industrial and Process Control, May*, 55–58.

Dodge, H. F. (1943). A sampling inspection plan for continuous production. *Annals of Mathematical Statistics, 14*, 264–279.

Jackson, J. E. (1972). All count distributions are not alike. *Journal of Quality Technology*, *4*, 86–92.

Ladwig, R. (1986). Modern data loggers: Same look, better machines. *Industrial and Process Control, May*, 51–53.

Page, E. S. (1954). Continuous inspection schemes. *Biometrika*, *41*, 101–115.

Wald, A. (1947). *Sequential Analysis.* Wiley, New York.

3
Implementation Strategies for On-Line Statistical Process Control Systems

Luis E. Contreras
ProScan, Inc., Austin, Texas

The successful implementation of on-line statistical process control systems requires a careful strategy to blend statistical procedures with hardware and software tools for data collection, analysis, and presentation. This paper describes findings on the current practice of SPC in manufacturing organizations, and presents an evolutionary "middle-out" implementation strategy with a discussion of the process of organizational change.

1. INTRODUCTION

Continuous flow manufacturing and other manufacturing policies based on advanced measurement, sensing, testing, and automation technologies generate very large volumes of process control data

Based on a presentation made at "Statistical Process Control: Keeping Pace with Automated Manufacturing, a National Symposium," sponsored by the Center for Professional Development and the Reliability, Availability and Serviceability Laboratory, College of Engineering and Applied Sciences, Arizona State University, November 6–7, 1986.

that must be stored, managed, analyzed, and presented with an *on-line statistical process control system* (OLSPCS).

The implementation of an OLSPCS is a complex process involving many technical and managerial considerations. It requires a careful strategy to blend statistical procedures with hardware and software tools for data collection, analysis, and presentation.

There are two extreme approaches to process control: centralized versus distributed control. Centralized control relies on the concept of a "control room" with access to all of the data in the manufacturing floor. In this scenario decisions are made at the top once the necessary data has been collected and organized for analysis. Distributed control, on the other hand, is based on the assumption that the line operator is qualified to take corrective actions and that corrections are required as soon as possible. Personal computer (PC) based workstations along with new hardware and software tools make distributed control a feasible alternative.

The *middle-out* implementation strategy described herein addresses the key implementation issues with an emphasis on the process of organizational change.

2. ON-LINE STATISTICAL PROCESS CONTROL SYSTEMS

Quality control involves many decisions and decision makers, each requiring different data and tools. The spectrum of decisions includes the following:

Decisions	Decision makers
Set quality goals	Plant manager
Select suppliers	Procurement manager
Design products	Production engineering manager
Develop processes	Process development engineer

Review designs	Manufacturing engineering manager
Evaluate process changes	Manufacturing engineer
Define assurance policies	Quality control manager
Analyze quality problems	Quality engineer
Control daily yields	Production manager
Adjust process performance	Line supervisor
Monitor process performance	Line operator

An OLSPCS supports directly the on-line analysis of incoming data to determine whether process modifications are in order. This requires the manipulation of large volumes of data for one or more signals and a simple display for the status of the process. Ideally, the same system with a different configuration supports the remaining quality decisions. Designing an OLSPCS requires defining the following issues:

Which processes should be controlled?
Who has the responsibility and authority?
Which are the relevant factors?
How do you measure these factors?
What are the specifications for these factors?
How do you collect the data?
What statistical control schemes apply?

3. HARDWARE AND SOFTWARE REQUIREMENTS

A process is either in control or out of control, and within specifications or out of specifications. When a process is in control it may eventually go out of control. Once the process is out of control it will remain in this state until a corrective action is taken.

The objective of OLSPCS is to detect out-of-control conditions as soon as possible. Real process improvement requires in many cases process modifications as suggested by process optimization studies based on experimental design and response surface techniques. The diagram in Figure 1 illustrates the stages in process

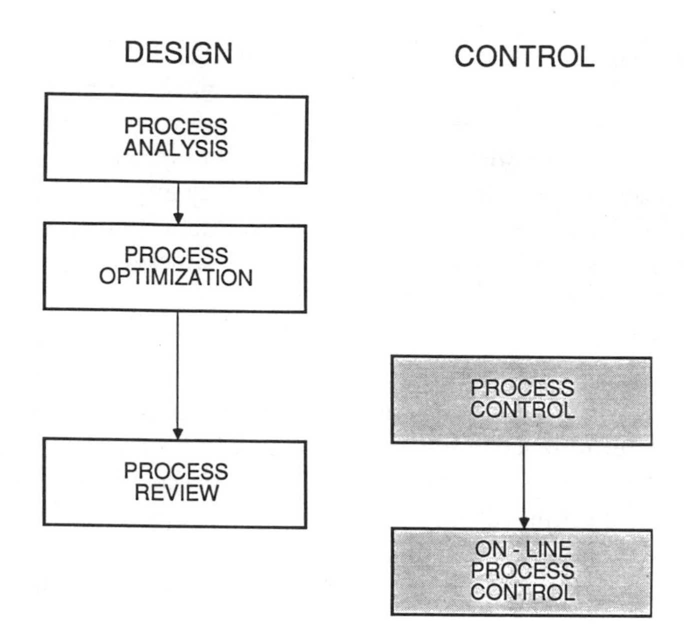

FIGURE 1 Process design and control.

design and control. Typically, manual process control procedures are installed to design and test OLSPCS.

In contrast with other information systems that require data from many points in a production line, process control data is distributed—that is, it originates at the point where it is used. While testing equipment and other sensors generate large volumes of data, only a fraction of that is relevant for OLSPCS. Historical data can help identify trends and cycles in a retrospective analysis, but in general it has a short useful life.

Figure 2 shows the layers of dependency of information in a process control system. *Measurement* is the specific hardware used to measure the relevant characteristics must be chosen according to the type of application. *Acquisition* is the process of collecting, sending, and receiving process control data from an on-line collection system with the required format translations. *Data management* is required since the volume and frequency of process control data varies over a wide range. It is important to store the data anticipating specific selections and extractions.

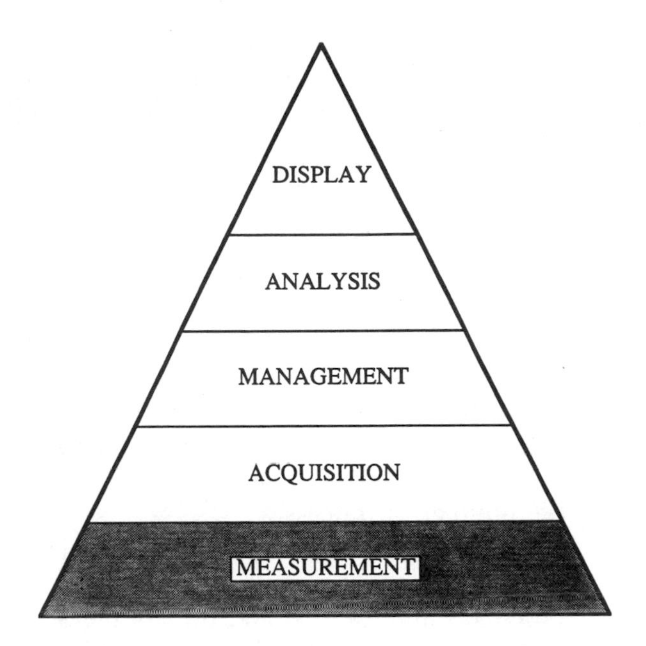

FIGURE 2 Process control information.

The following *analysis* features are frequently required for on-line applications:

Automatic detection of out-of-control conditions. Using pattern recognition and a set of user-defined decision rules, a system can detect when a process is going out of control. This feature provides "analysis by exception" where the operator displays control charts only when required.

Multidimensional analysis. A control chart tracks a single aspect of a quality characteristic, for example, slow drifts in the level of the process. To provide full control, multiple views or charts are often required.

Multiple signals. Process control data is frequently collected for multiple quality characteristics. These signals may be statistically independent but may be related in time or function. Thus, it is convenient to work with multiple signals at a time.

Multivariate analysis. If the quality characteristics describing a process are statistically dependent, multivariate process control techniques must be used. For this analysis it helps to study the univariate behavior with the component signals.

Autocorrelated data analysis. The usual assumption of independent observations within and between samples is violated in the case of on-line data collection where all units of production are measured. Several new techniques based on time series filters can be used to analyze on-line data.

The final aspect is *Display.* To be of any use the results of OLSPCS must be presented as charts or reports in a format that is adequate for the particular decision. A line operator needs a simple indication that the process must be explored and a reminder of the specific actions that he must take. On the other hand, a plant manager needs summary information for the behavior and capability of a large number of processes.

A brief comment on sampling may be useful at this point. There are many cases where sampling must be used. In addition to the case of destructive testing, the cost and time of measuring a given quality characteristic may be unacceptable for high-volume systems. For well-behaved systems the value of on-line process control information may be lower than its cost. For new processes and for the many engineering changes required in practice, sampling is the practical way to go. An OLSPCS software system should include these functions along with special on-line statistical control schemes.

4. THE IMPLEMENTATION PROCESS

To implement an OLSPCS the following questions need to be answered:

How does it start?

What resources are needed?

What are the expected benefits?

When do the activities take place?

Who is responsible for each task?

Who is responsible for each task?

How is the performance of the system measured?

To deal with these and other questions an implementation plan is needed that addresses many management, technical, and implementation issues.

4.1 Management Issues

The implementation and continued use of OLSPCS requires long-term commitments. Some of the common problems are:

Lack of real management commitment. Management must participate and lead the use of OLSPCS. It is not uncommon to find that process control data are collected and stored but not used to make process decisions, and that control charts are maintained but not acted upon. This is a common scenario when control charts are prepared to satisfy contractual requirements and not for process improvement.

Poor training. Engineers and other manufacturing personnel often ignore the variability in a process, and the impact on its yield. This often happens when the sources of variability are small, multiple, and far away from an inspection point where the yield is measured.

Reward systems. Management is driven by goals and targets on the yield of a process. The focus is on day-to-day fire fighting and away from real process improvement.

4.2 Technical Issues

OLSPCS often presents complex problems dealing with data measurement, acquisition, and analysis. In particular, one may encounter the following:

Data integration. The integration of different sources of data, frequencies, volumes, and formats often presents difficult technical challenges. Local area networks for manufacturing applications and other information systems developments are providing answers to these questions.

Measurement technologies. Recent development in measurement technologies provides efficient ways of measuring the desired quality characteristics.

Implementation Issues

While many companies talk about OLSPCS systems and are interested in their potential, only a handful of companies have implemented successful applications. There is a big gap between the theory and the application of these systems. Some of the implementation problems include the following:

Lack of definition. This problem manifests itself when different people have different ideas as to what the system is supposed to do.

Unrealistic expectations. After the initial success with a phase of the process, it is common to expect similar results in other areas, ignoring the differences in scale and applications.

Fragmentation of efforts. Without a sound implementation plan it is easy to start several phases of a project.

"*All or nothing.*" Requiring a comprehensive program before any stage may be attempted denies the benefits of improving certain key areas that may be responsible for a large number of quality problems.

"*Show me.*" A requirement to show tangible benefits before the system is installed.

"*Let's use our own.*" Manufacturing organizations with large investments in testing, data collection, computing, or other equipment often insist on using these resources, ignoring their limitations and the difficulties in communicating with other systems.

"*It may happen again.*" If a related project was unsuccessful, new efforts may be doomed from the beginning.

"*We tried it before.*" Previous experiences, perhaps under different circumstances and with different goals, resources, and tools, may hinder a revised project.

"*Once system X is installed.*" This is similar to the "all or nothing" objection and may delay a project indefinitely, since many other systems may be started along the way.

"*Ready by . . . .*" Rigid deadlines constrain the natural evolution of a process.

Counter–counter implementation. There are many ways of resisting change. In addition to straightforward denials, delays, postponements, etc., there are other subtle ways of resisting change.

One of the most interesting is known as "counter–counter implementation," where the scope of the project is expanded without increasing its schedule or resources. Typically this is done without the sponsor of the project being aware of the problem.

4.3 The Implementation Plan

To be successful in implementing an OLSPCS it is essential to prepare and use an implementation plan at the outset. This is a working document prepared and agreed to by all parties involved, defining the following:

The scope of the system
The calendar for the activities
The resources assigned
The review points
The expected results

5. MIDDLE-OUT IMPLEMENTATION STRATEGY

Traditional "bottom-up" and "top-down" implementation strategies are effective for static, well-defined projects. The middle-out implementation strategy is a sequential, iterative, evolutionary approach for complex and changing environments. It is based on small incremental changes with continuous planning and feedback, using a Master Plan as a reference. The emphasis is in the learning process and adaptation. Figure 3 illustrates the relationship of the sequential phases.

1. Release. To initiate the process of change, it is necessary to "unfreeze" the current conditions. Training seminars and other activities with top management commitment and participation are important at this stage.
2. Change. Choose a fast, visible, high-payoff application and find a "champion" to carry it out. Use a small team at low cost.
3. Freeze. This marks the completion of the application. The process is delivered to the unit in charge of day-to-day operations.
4. Review. Evaluate the changes made against the Master Plan and choose the next step.

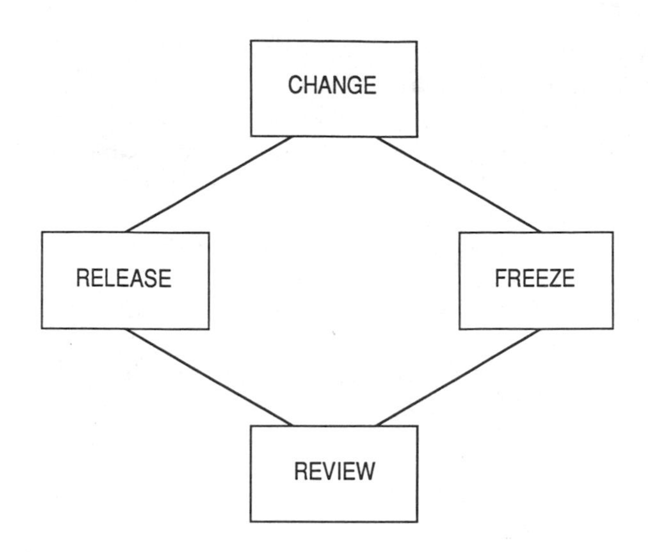

FIGURE 3 Middle-out implementation strategy.

6. CONCLUSIONS

The implementation of an OLSPCS is a complex process involving many technical and managerial considerations. It requires a careful strategy to blend statistical procedures with hardware and software tools for data collection, analysis and presentation.

The middle-out implementation strategy described herein addresses the key implementation issues with an emphasis on the process of organizational change. This strategy is a sequential, evolutionary approach for complex and changing manufacturing environments. It is based on small incremental changes with continuous planning and feedback.

SECTION II

Time Series Applications in Statistical Process Control

4
Time Series Modeling for Statistical Process Control

Layth C. Alwan and Harry V. Roberts
University of Chicago
Chicago, Illinois

In statistical process control, a state of statistical control is identified with a process generating *independent and identically distributed random variables* (IID). It is often difficult in practice to attain a state of statistical control in this strict sense; autocorrelations and other systematic time series effects are often substantial. In the face of these effects, standard control chart procedures can be seriously misleading. We propose and illustrate statistical modeling and fitting of time series effects and the application of standard control chart procedures to the residuals from these fits. The fitted values can be plotted separately to show estimates of the systematic effects.

1. INTRODUCTION

In standard applications of statistical process control, a state of statistical control is identified with a random process, that is, a process generating independent and identically distributed random variables (IID). Once a state of statistical control is attained, departures from statistical control may occur. These departures typically are reflected in extreme individual observations (outliers) or aberrant sequences of observations (runs above and below a level or runs up and down).

Departures from a state of statistical control are discovered by plotting and viewing data on a variety of control charts, such as Shewhart, *cumulative sum* (CUSUM), *exponentially weighted moving average* (EWMA), and moving-average charts. Having found departures, we hope to find explanations for them in terms of assignable or special causes. ["Assignable cause" is a term introduced by Shewhart (1931); "special cause" is an alternative term suggested by Deming (1982).] We then hope to move from "out-of-control" to "in-control" by correcting or removing the special causes.

In practice, however, it may be difficult either to recognize a state of statistical control or to identify departures from one because *systematic nonrandom patterns*—reflecting common causes—may be present throughout the data. [The term "common cause" was suggested by Deming (1982).] When systematic nonrandom patterns are present, casual inspection makes it hard to separate special causes and common causes.

A natural solution to this difficulty is to model systematic nonrandom patterns by time series models that go beyond the simple benchmark of IID. One possibility, for example, is a first-order autoregressive model, in which each observation may be regarded as having arisen from a regression model for which the current observation on the process is the dependent variable and the previous observation is the independent variable.

If such a time series model fits the data, leaving only residuals that are consistent with randomness, it is futile to search for departures from statistical control and their corresponding special causes. The practical emphasis would then shift to trying to gain better general understanding of the process. Process improvement would be sought by identification and understanding of the common causes making for autocorrelated behavior. Autocorrelated behavior means that there are carryover effects from earlier observations. The mechanism of these carryover effects must be sought.

Whether or not we achieve full understanding of the common causes underlying autoregressive behavior, fitting of the autoregressive model makes it possible, by study of residuals from the autoregressive model, to isolate the departures from control that may be traceable to special causes. Otherwise, these departures are confounded with the dominant autoregressive behavior of the data.

Hence, when the data suggest lack of statistical control, one should attempt to model systematic nonrandom behavior by time series models—autoregressive or other—*before* searching for special causes. In particular, we suggest using the ARIMA (autoregressive integrated moving average) models of Box and Jenkins (1970/1976), identified and estimated by standard techniques, to supplement the IID model as the guiding paradigm in practical work. This approach leads to two basic charts rather than one:

1. *Common cause chart: a chart of fitted values based on ARIMA models.* This chart provides guidance in seeking better understanding of the process and in achieving real-time process control.
2. *Special cause chart: a chart of residuals (or one-step prediction errors) from fitted ARIMA models.* This chart can be used in traditional ways to detect any special causes, without the danger of confounding of special causes with common causes. The IID model guides the interpretation of this second chart: all traditional tools of process control are applicable to it.

This strategy applies not only to industrial processes that manufacture items sequentially but also to quality control and statistical studies generally in all areas of businesses and other organizations.

In the light of the widespread use of ARIMA models in other areas of statistical application, we find it surprising that the practice of statistical process control has not moved in the direction here proposed. Although there have been many applications of time series concepts in process control, *the thrust of these applications has been directed toward testing for randomness, not toward modeling of departures from randomness.* In writings on quality control, we have been able to locate only one suggestion along the lines of this paper: Montgomery (1985, p. 265) outlines the strategy of basing control charts on residuals from time series modeling. In addition, a referee has pointed out an application in which standard control charting procedures were used by Berthouex et al. (1978) to study the residuals from a time series model.

The use of time series models requires more statistical skill than the use of traditional Shewhart charts, because one must know something about modeling nonrandom time series. But when the Shewhart charts are not fully pertinent to an application, there is little choice. Moreover, the level of skill needed to work with time series models is less demanding than at first appears. First, ARIMA modeling has to some extent been automated by computer programs. Second, a few simple special cases of ARIMA models, such as the first-order integrated moving average process—ARIMA(0, 1, 1)—may serve as good approximations for many or even most practical applications. [The EWMA chart is based on ARIMA(0, 1, 1).]

In addition, we believe that greater knowledge of time series methods would lead to more skilled interpretation of control chart data, even in the absence of formal time series modeling.

The plan of the paper is as follows. In Section 2 we discuss Shewhart's definition of statistical control and sketch its practical implementation. In Section 3, we explore limitations of the traditional implementation of process control. In Section 4, we outline our suggested extension of the traditional implementation. In Section 5, we fill in details of our proposals, offer an illustrative application, and briefly discuss typical applications. In Section 6, we consider possible ways of easing the demands for expertise in time series analysis required by our proposals. In Section 7, we outline various procedures required for full exploitation of our proposals, particularly for the interpretation of what lies behind systematic variation through time and what can be done to deal intelligently with this variation.

2. SHEWHART'S DEFINITION OF A STATE OF STATISTICAL CONTROL

Since the classic pioneering work of Walter A. Shewhart (1931)—*Economic Control of the Quality of Manufactured Product*—the concept of a state of statistical control has been central to prediction and control of industrial and other processes. Shewhart defined a state of control as follows: "a phenomenon will be said to be controlled when, through the use of past experience, we can predict, at least within limits, how the phenomenon may be expected to vary in the future. Here it is understood that prediction within limits means that we can state, at least approximately, the probability that the observed phenomcnon will fall within the given limits."

As implemented by Shewhart and his colleagues and successors, this definition of a state of statistical control has been specialized from general probabilistic prediction to prediction based on IID or simple random behavior. Further, in many applications the conditional distribution of individual quality measures, or of statistics computed from subsamples of these measures, can be approximated by familiar distributions such as the normal, binomial, and Poisson. Shewhart control charts based on these distributions became, and have remained, simple and powerful tools.

For example, a process yielding a quantitative quality measure can be regarded as in a state of statistical control if means of subsamples of five drawn at fixed time intervals behave as IID normal variables and if ranges of the same subsamples behave as IID variables following the distribution of ranges in samples of five from a normal distribution. In modern terminology, investigation of whether or not a process is in a state of statistical control means "diagnostic checking" of data from the process to see if the observed behavior is compatible with these specifications about the process. A Shewhart chart for variables facilitates this diagnostic checking, as do other common control charts.

If the verdict of checking is that the process is in a state of statistical control, there is need for surveillance to assure that the state of control continues or to sound an alert if it does not. If an alert is sounded, investigation and, possibly, corrective action are called for. One great advantage of the Shewhart chart is that the investigation for special causes is facilitated if undertaken soon after the alert is sounded.

Checking for a state of statistical control is usually regarded as a test of a null hypothesis. An out-of-control state is any hypothesis alternative to this null hypothesis. Procedures for practical process control draw heavily on hypothesis testing procedures suggested by statistical theory. For example, an inference of departure from control often is attached to a single subsample mean outside "three-sigma control limits" on a Shewhart control chart for means (XBAR chart)—roughly the 0.003 level of significance.

This particular test is reasonable if one believes that an important alternative hypothesis to the hypothesis of statistical control is the possibility that sudden and substantial shocks may impinge on the "constant cause" system underlying a process that has previously been in control. Against such an alternative hypothesis, the three-sigma control limits provide sufficient power to detect major shocks without sounding frequent false alarms in the absence of such shocks.

Shewhart applied the expression "assignable cause" to sources of sudden and substantial shocks, but he envisaged other kinds of assignable causes. For example, for economic data he mentioned "such things as trends, cycles, and seasonals." Most of the extensions of Shewhart's control chart procedures can be regarded as tests that are sensitive to particular departures from IID behavior. For example, there are criteria based on counts of runs above and below a given level, runs up and down, and runs of points within the three-sigma limits but more than, say, two standard deviations from the mean. Other tests are based on the cumulative sum chart, or CUSUM, originally introduced by Page (1954). There are also control charts based on moving averages and exponentially weighted moving averages (EWMA). For discussion of the various approaches to control charts see, for example, Montgomery (1985) or Wadsworth et al. (1986).

3. LIMITATIONS OF THE TRADITIONAL IMPLEMENTATION OF PROCESS CONTROL

Underlying standard control-chart procedures, there is a view of reality that envisages just two possibilities: a state of statistical control versus everything else. "A state of statistical control" is a sharp concept; "everything else" is not, although in particular applications it is sometimes made concrete in terms of specific alternatives against which one hopes tests to be powerful, such as sudden and substantial shocks or gradual trends.

Although this dichotomy of statistical control versus everything else often serves well in practice, it is not required by Shewhart's basic idea of a state of control, which requires only that "we can predict, at least within limits, how the phenomenon may be expected to vary in the future," and it may needlessly limit our perspective in applied work:

1. If one looks closely, "everything else" is likely to be the rule rather than the exception. A state of statistical control is often hard to attain; indeed, in many applications, it appears that this state is never achieved, except possibly as a very crude approximation. (A reader who doubts this will find it interesting—as one of us has done—to try to achieve a state of statistical control for body weight.) Our examination of published and unpublished data from actual applications suggests that distinctly nonrandom behavior will often be found in claimed examples of statistical control. In particular, substantial autoregressive behavior is very common.

   An interesting discussion of this point is provided by Churchill Eisenhart (1962, p. 167), who comments: "Experience shows that in the case of measurement processes the ideal of strict statistical control that Shewhart prescribes is usually very difficult to attain, just as in the case of industrial production processes. Indeed, many measurement processes simply do not and, it would seem, cannot be made to conform to this ideal...." Eisenhart goes on to cite an earlier comment of Student along the same lines.
2. An out-of-control process can be nonetheless a predictable process, one that is not affected by special causes. The failure to realize this, however, may lead people to view such a process as a sequence of isolated episodes, each with its own special cause and associated hint as to appropriate intervention. Preoccupation with nonexistent special causes diverts attention from common causes. Thus, if the data are autocorrelated, one may fail to look for factors that make for lagged effects.

Consider an application that we have examined recently, one that is far removed from industrial practice. The time series behavior of monthly closing of the Dow Jones Industrial Index from August 1968, through March 1986 is far from random. Technical analysts of the stock market often suggest special causes for variations of this index: for example, "profit taking," "concern about the budget deficit," or "lowering of the discount rate."

Yet the first differences of logs of the Dow Jones Index are nearly IID normal, to a closer approximation than we have typically seen in industrial data from processes alleged to be in control. For the stock market, then, a very simple application of time series modeling suggests that there are no special causes for market behavior in the sense of sudden shocks that are incompatible with a state of statistical control. Rather, we are observing a simple time series model, namely a *random walk* on the logs of the Index.

3. Emphasis on, say, a normal IID process as the model for a state of statistical control tends to encourage the development of a complex set of decision or testing procedures for detecting out-of-control situations. These rules may be based on combined consideration of individual extreme points, checks of runs above and below a given point and runs up and down, cumulative sum charts, etc. In testing language, we have multiple tests, so that the correct risks of Type I error must be calculated by some kind of compounding of probabilities, and this calculation, if possible at all, is very difficult.

   The focus on normal IID also tends to lead people to place undue emphasis on normality, since control limits are sometimes calculated without detailed scrutiny of the sequence plots of individual observations or even of control charts on means. Moreover, approximate normality of a histogram is often erroneously assumed to imply a state of statistical control for the process. For an example, see Deming (1982, p. 114).

4. SUGGESTED EXTENSION OF TRADITIONAL PROCESS CONTROL

In order to call attention to common causes when they are present, we suggest an alternative to the dichotomy of "a state of statistical control" versus "out-of-control." This alternative is based on the familiar decomposition of regression analysis:

ACTUAL = FITTED + RESIDUAL

(When the regression fit is used for forecasting, the parallel dichotomy is ACTUAL = PREDICTED + ERROR.)

Our experience suggests that in a wide range of applications in which processes are not in control in the sense of IID, one can use relatively elementary regression techniques to identify and fit appropriate time series models. If we succeed in finding such a model, we have reached a negative verdict about statistical control in the sense of IID, and we obtain fitted values and residuals along with probabilistic assessments of uncertainty. We can regard the process as "in control in a broader sense," a sense that is entirely consistent with Shewhart's conception of a state of statistical control.

Since successful time series modeling decomposes the actual series into fitted values and residuals, the traditional purposes of process control can be served by the two basic charts mentioned in Section 1:

1. A time series plot of the *fitted* values, without computation of control limits. This plot can be regarded as a series of point estimates of the conditional mean of a process—our best current guess based on past data of the location of the underlying process.
2. Standard control charts (Shewhart, CUSUM, or other) for the *residuals*. Control limits are based on the time series model itself; for example, limits for prediction errors would be based on the standard errors of one-step-ahead forecasts.

This two-step approach appears to be an obvious union of time series modeling with traditional ideas of process control, but the possibility of the union appears not to have been widely exploited. The closest approaches we have seen are those of Hoadley (1981) and J. S. Hunter (1986). So far as we can tell, other applications of time series concepts to process control have been oriented toward more sophisticated testing of the traditional null hypothesis of IID behavior. The power of proposed tests may be assessed against specific time series alternatives, but these alternatives are not explicitly modeled.

In the regression decomposition here proposed, each fitted value is conditioned only on past data rather than on all the data, as in signal extraction theory. We propose the regression decomposition because it articulates well with current statistical practice and can be implemented easily with standard computing packages.

5. BOX–JENKINS MODELING FOR PROCESS CONTROL

A useful set of tools for implementing these decompositions is provided by the modern principles of time series modeling, illustrated, for example, by another classic pioneering work, that of Box and Jenkins (1970/1976): *Time Series Analysis: Forecasting and Control.* Box and Jenkins stated their goals as follows:

> In this book we describe a statistical approach to forecasting time series and to the design of feed forward and feedback control schemes.... The control techniques discussed are closer to those of the control engineer than the standard quality control procedures developed by statisticians. This does not mean we believe that the traditional quality control chart is unimportant but rather than it performs a different function from that with which we are here concerned. An important function of standard control charts is to supply a continuous screening mechanism for detecting assignable causes of variation. Appropriate display of plant data ensures that changes that occur are quickly brought to the attention of those responsible for running the process. Knowing the answer to the question, "'when' did a change of this particular kind occur?" we can then ask "'why' did it occur?" Hence, a continuous incentive for process improvement, often leading to new thinking about the process, can be achieved.
>
> In contrast, the control schemes we discuss in this book are appropriate for the periodic, optimal adjustment of a manipulated variable, whose effect on some output quality characteristic is already known, so as to minimize the variation of that quality characteristic about some target value.
>
> The reason control is necessary is that there are inherent *disturbances* or *noise* in the process....

Our view is that Box–Jenkins time series methods—and their many extensions—are applicable both to the control schemes mentioned in this quotation *and* to the traditional process control objective of continuous process surveillance to detect special causes of variation. We believe that the basic philosophy of diagnostic checking espoused by Box and Jenkins leads almost inevitably to the approach to process surveillance outlined in this paper.

To illustrate time series modeling, we shall use a special case of Box–Jenkins ARIMA models known in statistical quality control as the exponentially weighted moving average or EWMA, explained in J. S. Hunter (1986). In Hunter's use of EWMA, EWMA is used as a testing procedure for testing a state of statistical control, although Hunter does emphasize the predictive quality of the EWMA, the need for prediction for control, and goes on to relate the EWMA to the control engineer's modeling approach. In our approach, the EWMA emerges as a flexible time series model that, for many but not all processes, may be a satisfactory approximation.

One appealing interpretation of EWMA is that the process being studied can be decomposed into two components:

1. IID random disturbances, mean zero
2. A *random walk* trend that is the sum of a fraction α of all past IID random disturbances

To bring out our main point, we shall consider an application in which individual observations rather than subsample means are used. We consider Series A of Box and Jenkins (1970/1976), 197 concentration readings on a chemical process, single readings taken every 2 hr. (We have subtracted a constant 10 from the actual readings.) The following computer output was produced by Minitab.

First, we show in Figure 1 the sequence plot of the series itself, called Y. From visual examination alone, we see below that the series is obviously out of control, with strong evidence of positively autocorrelated behavior. [See also Box and Jenkins (1970/1976, pp. 178–187).] The sample mean is 7.06 and the sample standard deviation is 0.40, so that conventional three-sigma control limits would extend from 5.86 to 8.26, limits that are shown in Figure 1

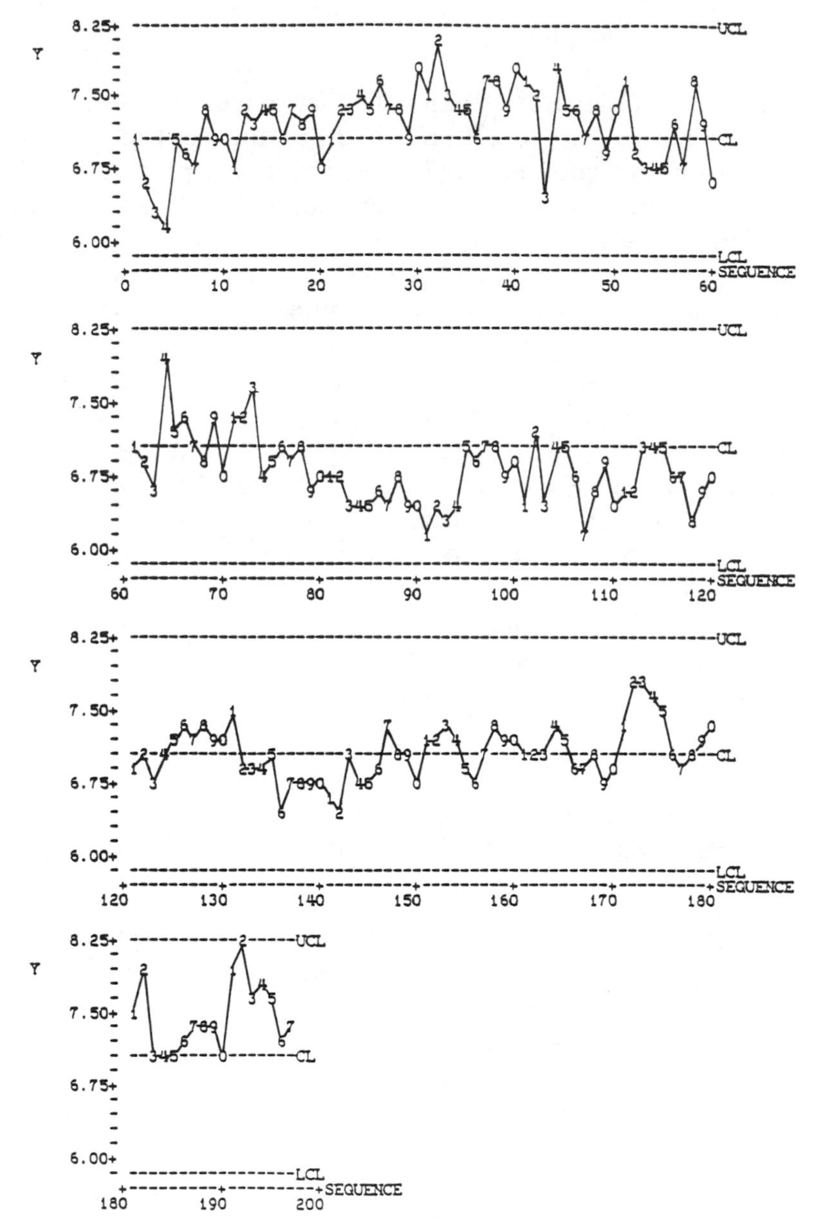

FIGURE 1 Sequence plot of Series A, Box and Jenkins (1970/1976): 197 concentration readings on a chemical process, single readings taken every 2 hr. Data are coded by subtraction of a constant 10 from each observation. CL is center line at height equal to sample mean, and UCL and LCL show location of conventional three-sigma control limits.

only for illustration. All points are within these limits (the maximum is 8.2 and the minimum is 6.1). (If the data are viewed without regard to time sequence, they conform very closely to the normal model, providing another illustration of nonrandom but approximately normal data.)

It is seen not only that the data are positively autocorrelated, but that it is not even obvious that the data should be regarded as coming from a stationary process. Even if the process were deemed stationary, however, three-sigma limits should logically be based on the *marginal* rather than the *conditional* standard deviation. However, any concern with control limits diverts attention from the basic observation that *the conditional mean is changing constantly.* These changes of the conditional mean are essential to understanding and control of the process.

A common approach for dealing with nonrandom data like these is to base control limits on the mean of moving ranges of successive observations (Wadsworth et al., 1986). We see in Figure 2 that these limits are much tighter and show many points out of control, occurring mainly at the peaks and troughs of the waves of the data.

These control limits based on moving ranges alert the user that the process is not close to being random, but they provide no real guidance in understanding what is happening.

For many users, data like these suggest a series of loosely connected episodes, each inviting ad hoc explanations. (This parallels the after-the-fact "explanations" offered by stock market analysts to "explain" changes of the Dow Jones Industrial Index!) The process of achieving a state of statistical control is sometimes pictured as finding special explanations for each episode, making a correction, observing the process further, finding further special explanations for further episodes, etc., until control is finally attained.

Here, however, as in the example of the Dow Jones Industrial Index, simple time series modeling unifies the picture immediately.

Box and Jenkins fit two models, each of which describes the data about equally well. We consider one of these models (the model underlying EWMA), namely, a first-order integrated moving average, called "ARIMA(0, 1, 1)," which, as pointed out above, can be interpreted as a random-walk trend plus a random deviation

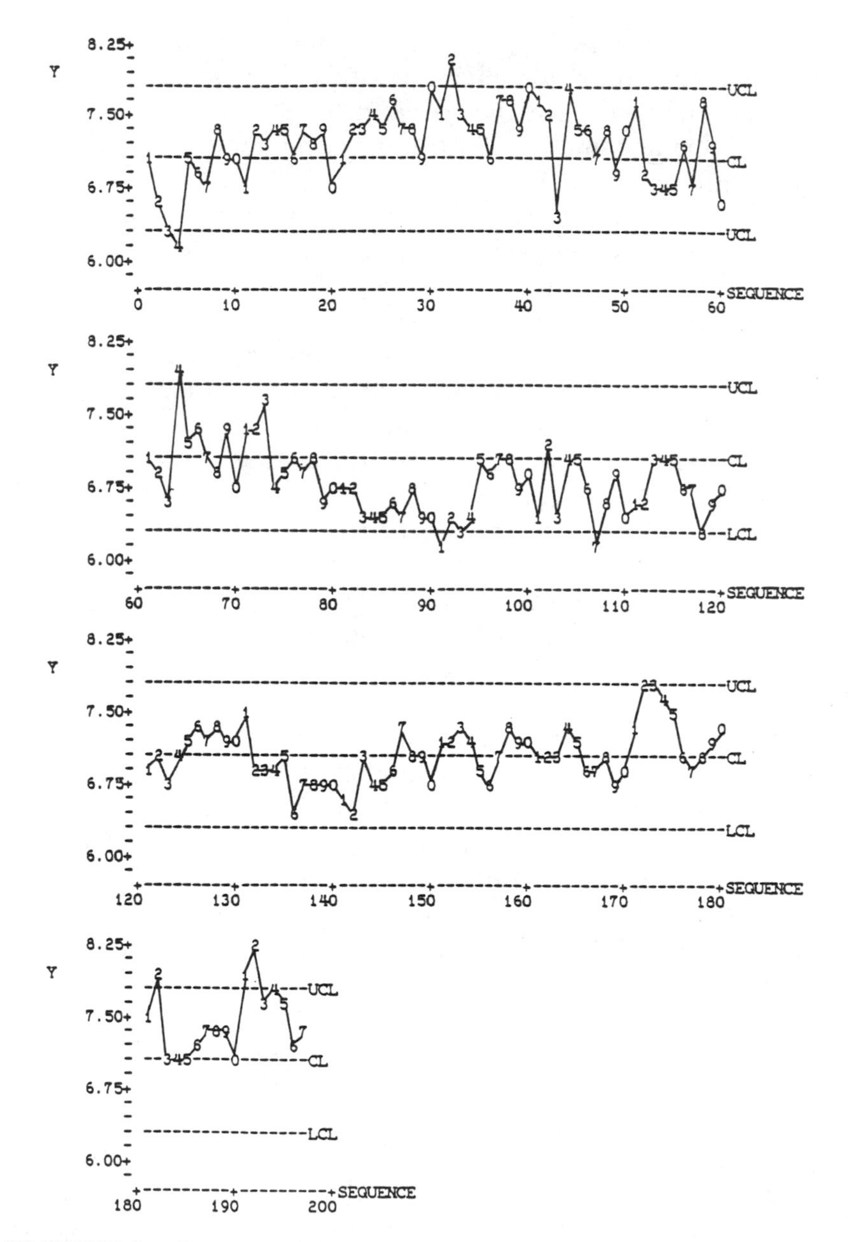

FIGURE 2 Sequence plot of same data as in Figure 1. UCL and LCL are limits computed from the mean of moving ranges of successive observations.

from trend. This model specifies that the observed Y_t is the sum of an unobserved random shock A_t plus a (proper) fraction α of the sum of all past random shocks $A_{t-1}, A_{t-2}, \ldots$. In the current application we have assumed that the mean of changes of Y_t is zero. As estimated by the Minitab ARIMA procedure, we have

$$\text{FITTED } Y_t = Y_{t-1} - 0.705A_{t-1}$$

The quantity 0.705 is an estimate of what is called θ by Box and Jenkins; $1 - 0.705 = 0.295$ is an estimate of what is often called α in discussions of the EWMA. The standard error of the estimated θ is 0.05 and the standard deviation of residuals is 0.318. (We have suppressed the constant term, for two reasons: we have judged such a term unlikely a priori and the sample evidence suggests an estimated constant near zero.)

5.1 Common Cause Chart: Fitted Values

It is useful to display the fitted values or estimated conditional means of the process, called FITTED in Figure 3, on a time sequence plot. This plot gives a view of the level of the process (estimated conditional mean) and of the evolution of that level through time. We see the systematic behavior of the process that pervades the entire period of observation. This behavior may aid in real-time control or in better understanding how the process is working.

The model just fitted can be interpreted as follows: each observation can be thought of as a random disturbance plus a random-walk trend or drift that reflects a certain fraction of the sum of all past random disturbances. Thus a part of each disturbance continues to affect the process into the indefinite future. The fitted values in Figure 3—FITTED—are estimates of the underlying random-walk trend: they follow a random walk without drift. The chart of FITTED represents the first of the two charts proposed by us. Each point is an estimate of the local level of the process itself, as distinguished from the observed readings.

To illustrate the type of control decisions that could be based on this plot, suppose that the most desirable level of the process is 7.0, and that increasing deviations from that level entail increasing

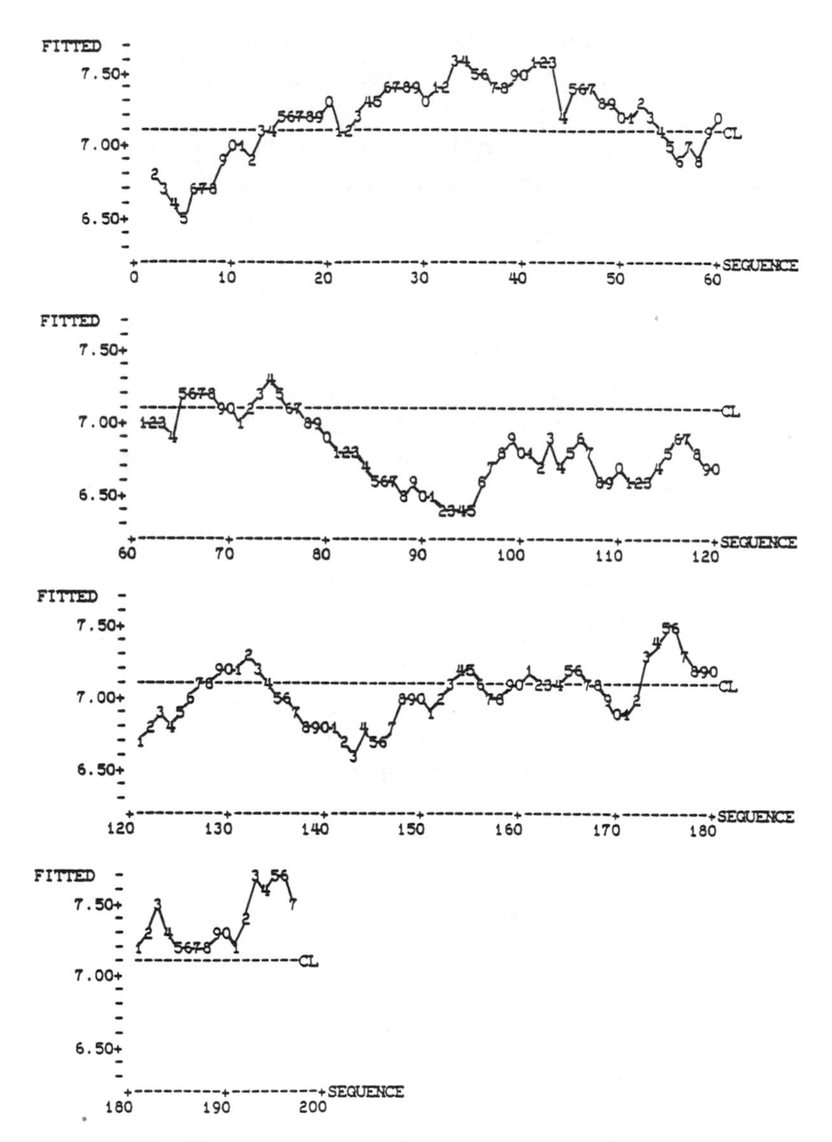

FIGURE 3 Common cause chart: sequence plot of fitted values for data of Figures 1 and 2. Estimates are based on the ARIMA(0, 1, 1) model with zero constant.

economic loss from less-than-optimal product. Suppose further that at a certain known cost it is possible at any time to re-center the process at 7.0. Then one can make an economic calculation to balance the expected loss of bad product over some specified period of time against the cost of re-centering. This calculation will define action limits both below and above 7.0 at which the process should be re-centered.

Note that these action limits are conceptually different from traditional control limits. They are not signals that it is time to look for special causes; rather, they are signals that a specific corrective action is needed.

5.2 Special Cause Chart: Residuals

The second graph is essentially a standard control chart for residuals. To facilitate visualization of control limits, we plot in Figure 4 the standardized residuals STRES, with approximatc center line and upper and lower three-sigma limits.

It is seen that two individual points—observations 43 and 64—breach the three-sigma limits. By contrast, the first set of control limits calculated previously suggested no points out of control, while the second set, based on mean ranges, suggested many individual points out of control. These out-of-control points did include observation 64, which was at the top of a wave, but not observation 43.

Also, the plot above reveals perhaps two or three short intervals for which runs counts would suggest some suspicion of lack of control. Note that runs counts are applied only after the residuals from the time series model have been isolated. As applied to the original series, runs counts would suggest that the process was nearly continually out of control, sometimes on the high side and sometimes on the low side.

5.3 Other Simple Models

The ARIMA(0, 1, 1) model fitted above is not the only reasonable model for the data. Box and Jenkins also fit ARIMA(1, 0, 1) with a nonzero constant. The two models give nearly identical fitted values and residuals, and hence about the same degree of overall fit. For

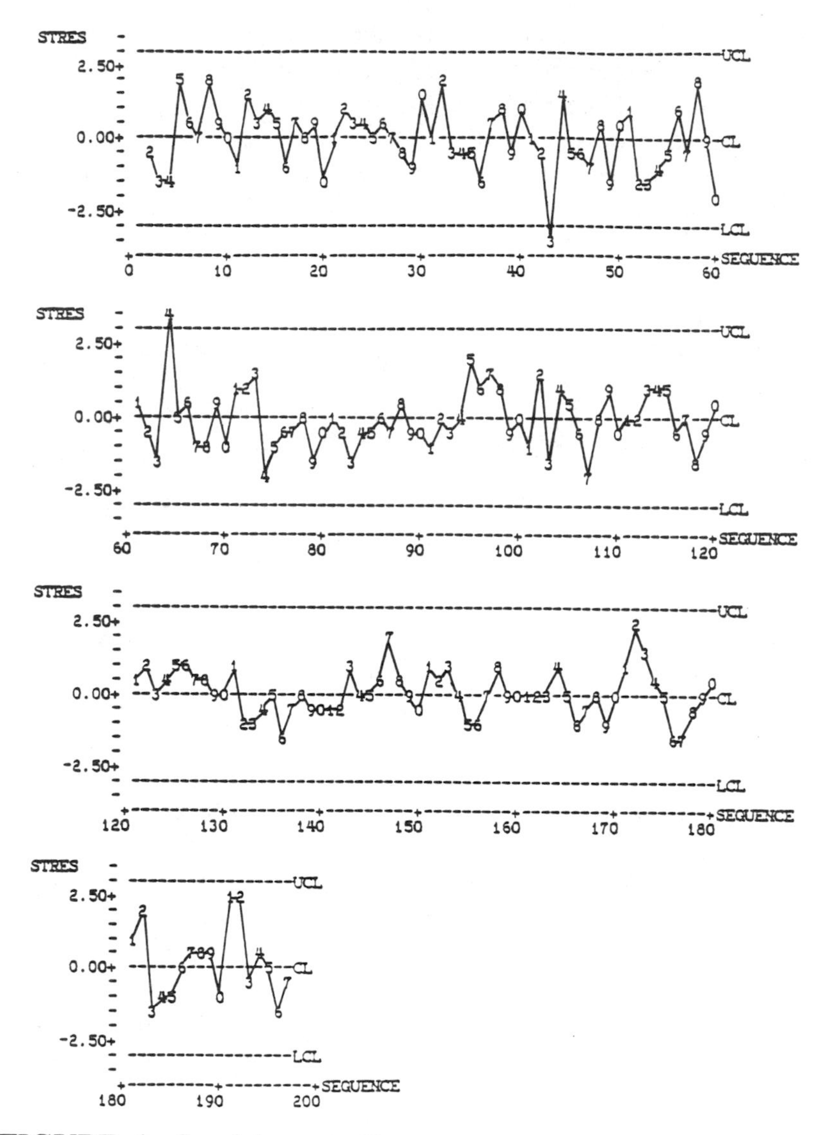

FIGURE 4 Special cause chart: sequence plot of residuals for data of Figures 1 and 2. Estimates are based on the ARIMA(0, 1, 1) model with zero constant.

many purposes of control, whether real-time control or process surveillance, either would be suitable.

From a theoretical point of view, the two models have very different implications for the long-run behavior of the process, assuming continuation of the basic conditions now being observed. ARIMA(0, 1, 1) is nonstationary, while ARIMA(1, 0, 1) is stationary. Given the ARIMA(0, 1, 1) fitted above, there is no tendency of the process to revert to its mean level as it wanders away. Given the ARIMA(1, 0, 1) for the same process, there is a relatively weak tendency to mean reversion. Moreover, prospective control limits as we look farther into the future are constant for ARIMA(1, 0, 1) but ever-widening for ARIMA(0, 1, 1).

At one time we contemplated the concept of a *stationary* process as a possible extension of the concept of IID in defining a state of statistical control. The present example illustrates, however, that data may not permit a sharp distinction between stationary and nonstationary. Hence we propose only that *some* time series model be used to define statistical control in an extended sense. Even a nonstationary model permits probabilistic prediction.

Of course, if the weight of statistical evidence and background knowledge suggests that the process should be regarded as nonstationary, the problem confronting management is inherently more difficult, since a nonstationary process has no tendency toward mean reversion. For example, a stationary ARIMA model about, say, a linear trend suggests disaster in the absence of intervention. At the same time, the chart of residuals could suggest that special causes are occurring along the road to ruin of the process itself.

6. TIME SERIES MODELING IN PRACTICE

A practical limitation on the use of the approach advocated in this paper is that implementation requires some skill in analysis of time series, whereas the implementation of the standard Shewhart procedures entails only the most elementary statistical knowledge. We believe that in many applications, the ability to better sort out special causes from common causes justifies the use of the more elaborate machinery required by our approach.

Although there can be no completely satisfactory substitute for statistical skills in time series analysis, we believe that much can be accomplished by people with limited skills. For example, modern computational tools make possible relatively automated implementation of time-series modeling; automatic fitting of Box–Jenkins models has for some time been available in software of general purpose software packages like Minitab, Statgraphics, SCA, or SYSTAT now available on the IBM PC (personal computer) and compatibles. (There are even programs that attempt to automate model identification as well as fitting! See, e.g., Shumway, 1986).

It is reassuring to know that precise model identification may not be essential to effective process control: several alternative models may fit the past data, at least, about equally well. In particular, the two models mentioned in our example in Section 5—ARIMA(1, 0, 1) and ARIMA(0, 1, 1)—offer reasonably good fit for a wide range of applications.

If a process can be modeled, then the traditional objectives of quality control—or surveillance—can be better served. Moreover, a control mechanism can be designed, and if a principal disturbance to the process can also be modeled, we have the adaptive control situation described in Box and Jenkins (1970/1976).

7. ENGINEERING AND ECONOMIC ISSUES

More difficult than model identification and fitting is the interpretation and exploitation of fitted values for real-time process control, especially when process re-centering is costly, and for better understanding of underlying common causes that make the process behave as it does. Real-time process control has received considerable attention, for example, by Box and Jenkins (1970) and Box et al. (1974). Systematic study of common causes has attracted little attention. We hope, in future work, to examine questions like these in some detail.

REFERENCES

Berthouex, P. M., W. G. Hunter, and L. Pallesen. (1978). Monitoring sewage treatment plants: Some quality control aspedts. *Journal of Quality Technology*, *10*(4), 139–149.

Box, G. E. P., and G. M. Jenkins. (1970/1976). *Time Series Analysis, Forecasting and Control.* Holden-Day, San Francisco.

Box, G. E. P., G. M. Jenkins, and J. F. MacGregor. (1974). Some recent advances in forecasting and control: Part II. *Applied Statistics, 23,* 158–179.

Deming, W. E. (1982). *Quality, Productivity and Competitive Position.* Cambridge, MA: MIT Center for Advanced Engineering Study, Cambridge, Mass.

Eisenhart, C. (1963). Realistic evaluation of the precision and accuracy of instrument calibration systems. *Journal of Research of the National Bureau of Standards—C. Engineering and Instrumentation, 67C,* 2.

Hoadley, B. (1981). The quality measurement plan. *Bell System Technical Journal, 60,* 215–271.

Hunter, J. S. (1986). The EWMA. *Journal of Quality Technology, 18,* 4.

Montgomery, D. C. (1968). *Introduction to Statistical Quality Control.* Wiley, New York.

Page, E. S. (1954). Continuous inspection schemes. *Biometrika, 41,* 100–115.

Shewhart, W. A. (1931). *Economic Control of Quality of Manufactured Product.* Van Nostrand, New York. Republished in 1981, with a dedication by W. Edwards Deming, by the American Society for Quality Control, Milwaukee, Wis.

Shumway, R. H. (1986). "AUTOBOX (Version 1.02)," *The American Statistician, 40,* 299–300.

Wadsworth, H. M., K. S. Stephens, and A. B. Godfrey. (1986). *Modern Methods for Quality Control and Improvement.* Wiley, New York.

5
Statistical Process Control in a Computer-Integrated Manufacturing Environment

Douglas C. Montgomery
University of Washington
Seattle, Washington

David J. Friedman
AT&T Engineering Research Center
Murray Hill, New Jersey

Statistical process control methods have been used in industry for over 50 years. In applying the usual Shewhart control charts, it is customary to assume that the observations on the process constitute a random sample; that is, the observations are uncorrelated with each other both between and within samples. In computer-integrated manufacturing, sensors are often used for real-time, online data capture. In such processes, observations may exhibit serial correlation. This paper discusses how serial correlation impacts the use of Shewhart, cumulative sum, and geometric moving average control charts. It is shown how these control chart procedures can be suitably modified for use with correlated data. The modification consists of modeling the original signal with an adequate stochastic model and then plotting the residuals from the model on a control

---

Based on a presentation made at "Statistical Process Control: Keeping Pace with Automated Manufacturing, a National Symposium," sponsored by the Center for Professional Development and the Reliability, Availability and Serviceability Laboratory, College of Engineering and Applied Sciences, Arizona State University, November 6–7, 1986.

chart. The procedure is illustrated for both univariate and multivariate data. Some suggestions for design of appropriate control schemes are given.

1. INTRODUCTION

Statistical process control methods have been used extensively for process monitoring and improvement, particularly in manufacturing industries. We typically justify their use by appealing to several assumptions. The most common and straightforward of these are as follows:

1. The data is obtained from the process via periodic samples.
2. Observations are statistically independent, both between and within samples.
3. Rational subgrouping is used in the selection of samples, and sample sizes are usually greater than one.
4. The data follows some particular probability distribution; for instance, with variables data it is customary to assume normality.

Since control charts are widely used in practice, and usually provide excellent results, we conclude that either the assumptions stated above are frequently satisfied, or that control charts are reasonably robust to moderate departures from some of these assumptions.

In this paper, we describe the use of statistical process controls in the *computer-integrated manufacturing* (CIM) environment. In such a situation sensors are frequently used to measure and capture functional parameter or quality data on every unit produced. Thus the assumption of uncorrelated data may no longer be reasonable. It is well known that the assumption of uncorrelated data is critical to the proper function of a control chart; for example, see Johnson and Bagshaw (1974), Bagshaw and Johnson (1975), and Vasilopoulus and Stamboulis (1978). Consequently, some modification to the usual procedure must be applied.

In this paper we present a general approach to analyzing correlated data, such as typically obtained in computer-integrated manufacturing, via control charts. The procedure consists of first modeling the process data with an appropriate empirical stochastic model, and then applying control charts to the residuals. We investigate three particular types of control charts: a Shewhart-type control chart for individual measurements, the cumulative sum control chart, and the geometric moving average control chart. We show that the cumulative sum and geometric moving average control charts are more effective in detecting small disturbances in the process. Recommendations for design of a control system for use in practice are also given.

The general idea of filtering a correlated signal to produce an uncorrelated sequence of residuals that can be analyzed by conventional control charts has been suggested by several other authors, although not in the computer-integrated manufacturing environment. Berthouex et al. (1978) illustrates how the general procedure could be applied to monitoring sewage treatment plant data. Of related interest are the papers by Notohardjono and Ermer (1986), Ermer (1980), and Liao et al. (1982).

2. STATISTICAL PROCESS CONTROL IN COMPUTER-INTEGRATED MANUFACTURING

The computer-integrated manufacturing environment typically includes automated or semiautomated machine processing centers and a material handling systems with at least partial computer control of schedule, material flow or parts movement, and quality information. It is not unusual to find sensors employed for on-line, real-time evaluation and measurement of functional parameters and critical quality characteristics in manufactured parts. Sensors allow immediate data capture for these functional parameters on every unit produced. A schematic representation of this system is shown in Figure 1. Thus, in typical computer-integrated manufacturing environments, the sample size for surveillance and control purposes is $n = 1$. This is in marked contrast with the rational subgrouping approach, which for measurements data typically leads to periodic samples with sample sizes greater than one. Because measurements are made on each successive part, the assumption of uncorrelated measurements may not be realistic for most computer-integrated manufacturing process control problems. Consequently, one may seriously doubt whether a standard

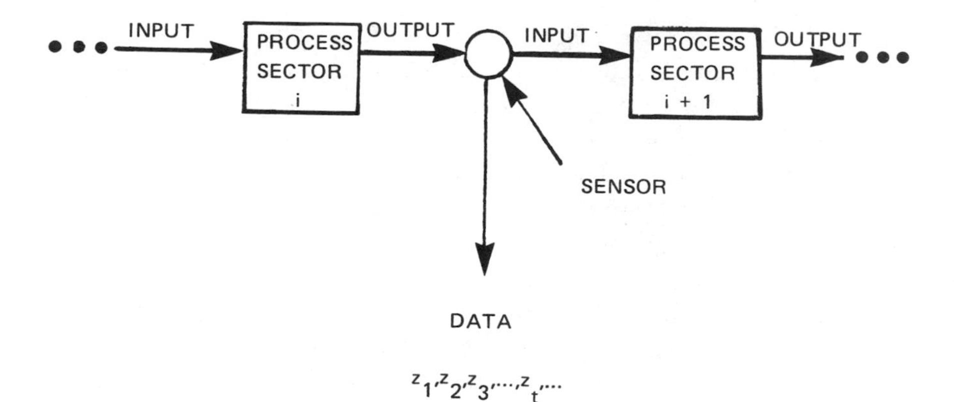

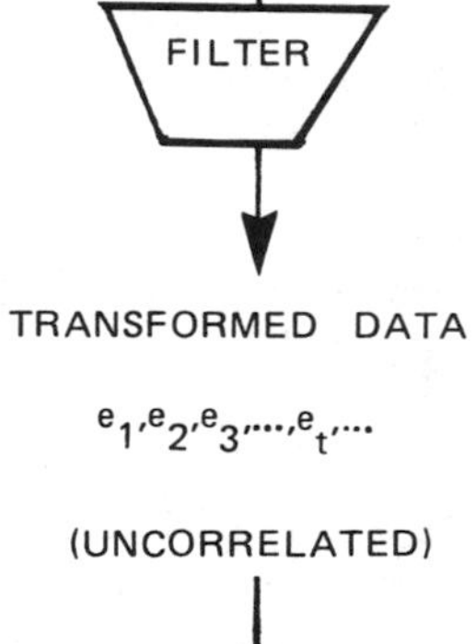

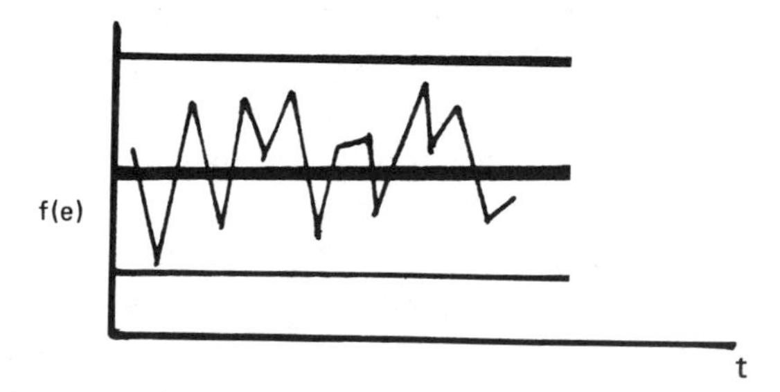

FIGURE 1 SPC system for computer-integrated manufacturing environment.

control chart will work satisfactorily for data from such a manufacturing process. As noted previously, control charts are not robust with respect to departures from the independent or uncorrelated data assumption. Other assumptions, such as normality, are of somewhat less concern.

One method for dealing with serial correlation is to model the original process data with a suitable empirical stochastic model. The residuals from such a model would then be uncorrelated if the process is in statistical control. The usual control charting methods of statistical process control could then be applied to the residuals. Changes in either the level or dispersion of the original signal would then cause corresponding changes in the residuals, which would be detected on the standard control charts. Figure 1 shows schematically how such a system would operate. We now turn to the choice of appropriate stochastic models for the original process, and illustrate how various types of control charts perform when applied to the residuals.

3. DATA MODELING AND CONTROL CHARTS

In Figure 1, the original data from the process $Z_1, Z_2, Z_3, \ldots, Z_t, \ldots$, which may exhibit autocorrelative behavior, is transformed via a filter into an uncorrelated or white noise signal $e_1, e_2, e_3, \ldots, e_t, \ldots$, called the residuals. Then an appropriate function of the residuals, say $f(e)$, is plotted on a control chart. The *autoregressive integrated moving average* or ARMA (p, q) model is ideally suited for developing an empirical stochastic time series model of the original data. This model can accommodate not only correlated data structure, but nonstationary behavior, deterministic trends, interventions, and other disturbances.

The general form of the ARMA (p, q) model is

$$Z_t = \xi + \phi_1 Z_{t-1} + \phi_2 Z_{t-2} + \cdots + \phi_p Z_{t-p} + \varepsilon_t - \theta_1 \varepsilon_{t-1} - \theta_2 \varepsilon_{t-2} - \cdots - \theta_q \varepsilon_{t-q} \quad (1)$$

where $\phi_1, \phi_2, \ldots, \phi_p$ are the autoregressive parameters, $\theta_1, \theta_2, \ldots, \theta_q$ are the moving average parameters, and ξ is a parameter related to the level or mean of the series. If the original signal exhibits nonstationary behavior, then Z_t may be a suitably differenced signal. We may rewrite Eq. (1) as

$$\hat{\Phi}_p(B)\tilde{Z}_t = \varepsilon_t \Theta_q(B) \tag{2}$$

where $\hat{\Phi}_p(B) = (1 - \sum_{j=1}^{p} \phi_j B^j)$, $\Theta_q(B) = (1 - \sum_{j=1}^{q} \theta_j B^j)$, B is the usual backward difference operator $B^j Z_t = Z_{t-j}$, and $\tilde{Z}_t = Z_t - \bar{Z}$. Nonlinear least-squares methods may be used to estimate the parameters ϕ_j and θ_j.

Let $\hat{\Phi}_p(B)$ and $\hat{\Theta}_q(B)$ represent the autoregressive and moving average polynomials, respectively, with the parameters ϕ_j and θ_j replaced by their least-squares estimates. Then the residuals are given by

$$e_t = \hat{\Theta}_q^{-1}(B)\hat{\Phi}_p(B)\tilde{Z}_t \tag{3}$$

If the form of the empirical model of Eq. (2) is correct and the process is in-control, then the residuals e_t are uncorrelated.

There are several obvious candidates for control charts for the residuals. Since the sample size $n = 1$, the Shewhart-type chart would be the control chart for individuals and the moving range chart. The control chart for individuals would be constructed by plotting the residuals e_t on a control chart with parameters

$$\begin{aligned} \text{UCL} &= \bar{e} + \frac{3}{1.128}\bar{R}_m \\ \text{CL} &= \bar{e}\,(\cong 0) \\ \text{LCL} &= \bar{e} - \frac{3}{1.128}\bar{R}_m \end{aligned} \tag{4}$$

where $\bar{R}_m$ is the average of the moving ranges of successive pairs of time-ordered residuals. Note that the center line $\bar{e}$ should be approximately zero. The moving range control chart has the following parameters:

$$\begin{aligned} \text{UCL} &= 3.267\bar{R}_m \\ \text{CL} &= \bar{R}_m \\ \text{LCL} &= 0 \end{aligned} \tag{5}$$

Details of the construction and interpretation of these control charts are in Montgomery (1985, Ch. 6).

A two-sided *cumulative sum* (CUSUM) control chart could also be applied to the residuals. We suggest using the tabular form of the CUSUM rather than the more traditional V-mask scheme. This involves plotting two one-sided CUSUMS,

$$S_{\mathrm{H}}(t) = \max\,[0, e_t - (\bar{e} + k) + S_{\mathrm{H}}(t-1)] \tag{6a}$$

$$S_{\mathrm{L}}(t) = \max\,[0, (\bar{e} - k) - e_t + S_{\mathrm{L}}(t-1)] \tag{6b}$$

and concluding that the process is out of control if either $S_{\mathrm{H}}(t) > h$ or $S_{\mathrm{L}}(t) > h$. The CUSUM is designed by choosing the reference value k and the decision interval h. These parameters may be chosen to give specified average run length performance. See Lucas (1976) or Montgomery (1985, Ch. 7).

A useful alternative to the CUSUM is the *geometric moving average* (GMA) control chart. To construct this chart, plot the geometric moving average

$$G_t = re_t + (1-r)G_{t-1} \tag{7}$$

against the control chart parameters

$$\begin{aligned}
\mathrm{UCL} &= \bar{e} + \Delta \frac{\bar{R}_m}{1.128} \sqrt{\frac{r}{(2-r)}\,[1-(1-r)^{2t}]} \\
\mathrm{CL} &= \bar{e}\,(\cong 0) \\
\mathrm{LCL} &= \bar{e} - \Delta \frac{\bar{R}_m}{1.128} \sqrt{\frac{r}{(2-r)}\,[1-(-r)^{2t}]}
\end{aligned} \tag{8}$$

where $r\,(0 \leqslant r \leqslant 1)$ is a discount factor and Δ is the width of the control limits in chart-sigma units. Notice that the control limits will increase in width fairly rapidly and then stabilize, as the term $[1-(1-r)^{2t}]$ approaches unity.

If properly designed, the GMA will have average run length performance equivalent to the CUSUM while still preserving the general appearance of a Shewhart-type control chart. We usually recommend small values of r, say, $0.05 \leqslant r \leqslant 0.20$, and either $\Delta = 2.5$ or $\Delta = 3.0$. Further details of this control chart are in Montgomery (1985, Ch. 7) and Hunter (1986).

3. A UNIVARIATE EXAMPLE

To illustrate the procedure, we consider an application to a machining process in which a cable harness hole is drilled in a wing leading edge rib. The diameter of the hole is a critical quality characteristic. Measurements are made on every hole drilled, in time order of production. The first 50 observations are taken during a time period when the process is thought to be in statistical control. These observations are shown to the left of the vertical line in Figure 2. Based on this sample of 50 measurements, a reasonable stochastic model of the process is

$$Z_t = 50 + 0.8Z_{t-1} - 0.3Z_{t-2} + \varepsilon_t \quad (9)$$

The second half of Figure 2 plots the next 50 observations taken from the process. There is a distinct visual impression of a downward shift in the process mean. Figure 2 also plots the control limits for a control chart for individual measurements, where the

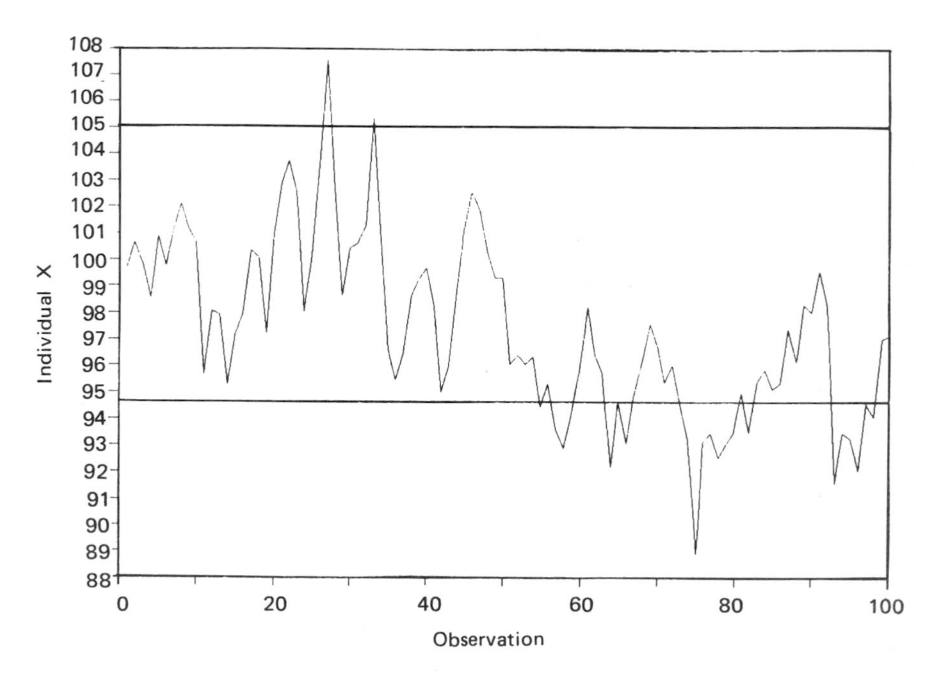

FIGURE 2 Control chart for individuals, original observations.

control limits have been computed from the original 50 observations. We immediately notice a serious flaw with this chart: during the original 50 observations, when the process was thought to be in-control, several out-of-control signals are generated. In the latter half of the series, where we suspect that a shift in the mean has occurred, the control chart does indicate a downward shift in the mean since there are several points below the lower three-sigma control limits. Warning limits, run tests, or other sensitizing procedures could also be applied to the control chart so that small shifts could be detected more quickly.

The flaw highlighted on this chart occurs repeatedly with correlated data. Generally, the effect of serial correlation is to cause too many false out-of-control signals. False alarms are a serious occurrence in statistical process control, and their effect in the CIM environment could potentially be disastrous.

Figures 3 and 4 present the CUSUM and GMA control charts applied to the original process data. Notice that these charts also

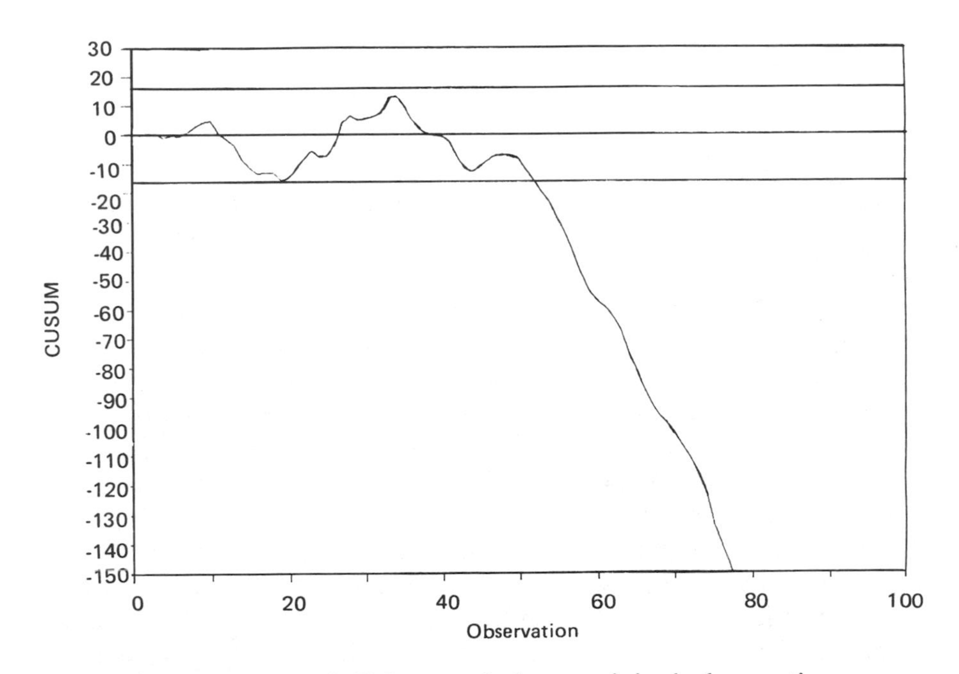

FIGURE 3 CUSUM control chart, original observations.

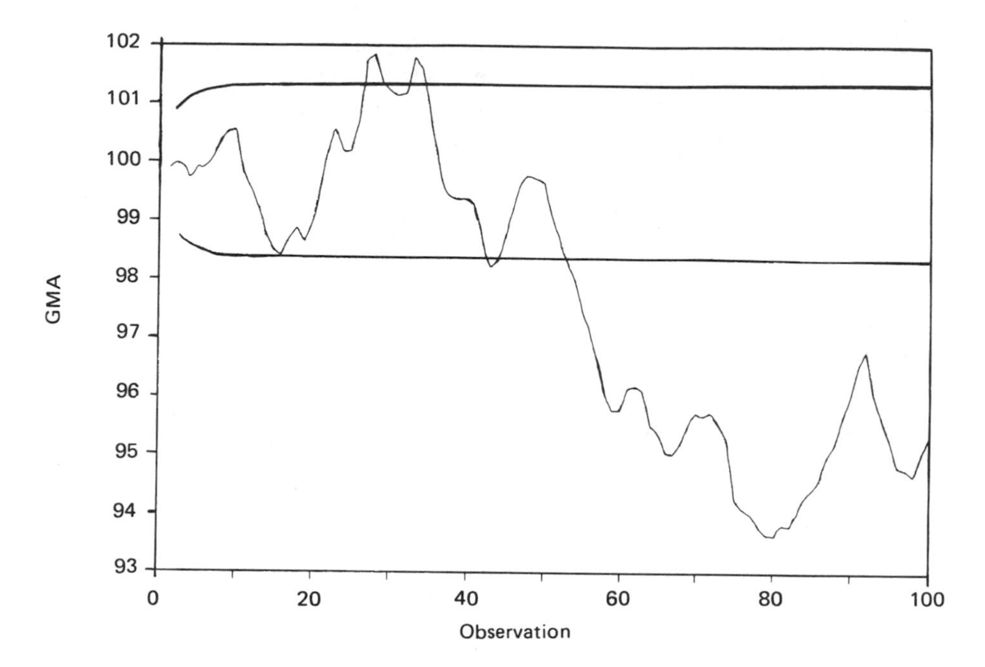

FIGURE 4 GMA control chart, original observations.

suffer from too many false alarms during the period when the process is thought to be in control. However, they are more effective in detecting the apparent shift in the process mean that occurs around observation 50.

Figure 5 presents the control chart for individuals applied to the residuals from the model [Eq. (9)]. Notice that the first 50 observations come from a process that is in statistical control, confirming our original hypothesis. The pattern generated on the second half of the chart is distinctly that of a process for which the mean has shifted downward, and, eventually, points fall below the lower control limit. The use of warning limits and other sensitizing rules would be required to provide an early detection of the shift.

Figures 6 and 7 present the CUSUM and GMA control charts for the residual values. Notice that these charts also indicate that the first 50 observations are in control. However, they respond much more quickly to the out-of-control condition in the second

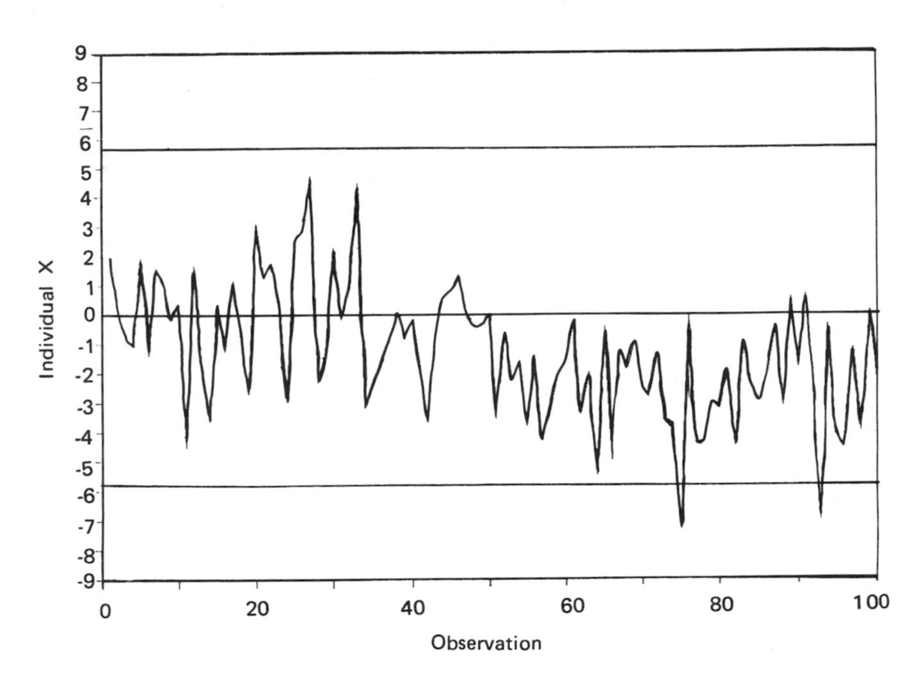

FIGURE 5 Control chart for individuals, residual values.

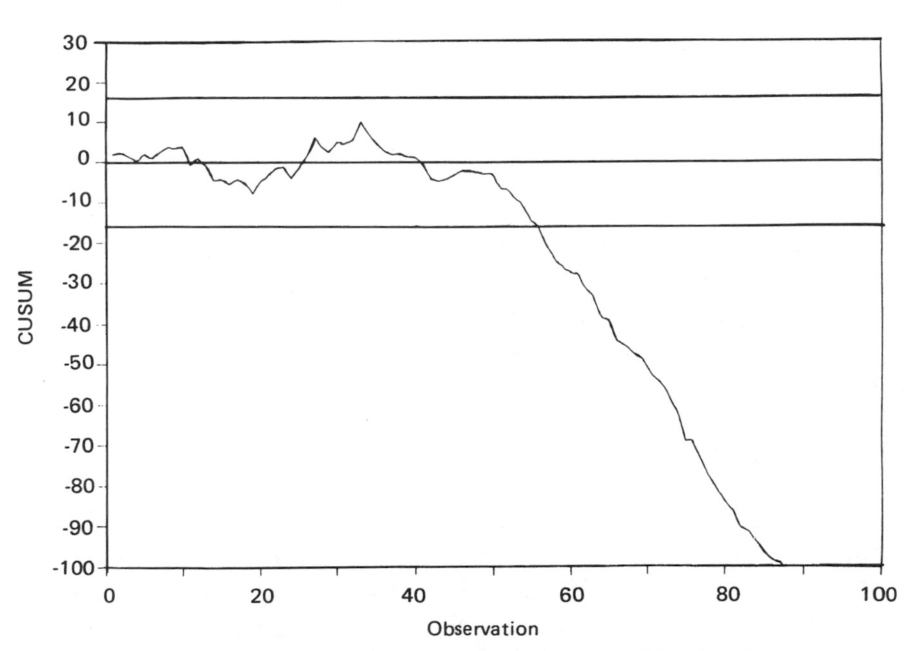

FIGURE 6 CUSUM control chart, residual values.

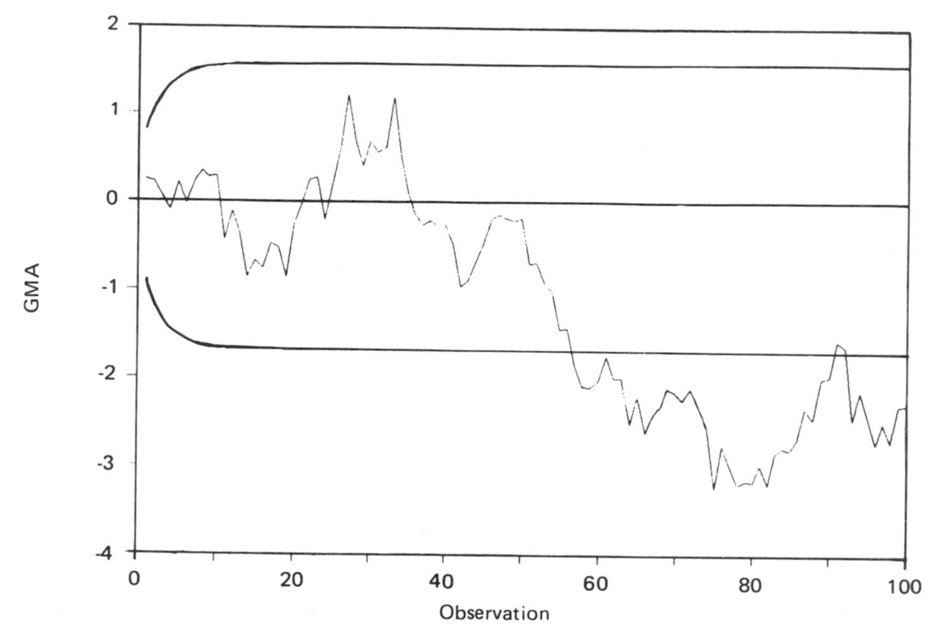

FIGURE 7 GMA control chart, residual values.

half of the series than does the control chart for individual measurements. This will almost always occur for small process shifts, as both the GMA and CUSUM control charts are more effective than their Shewhart counterparts for small shifts. Large shifts may be more rapidly detected on a Shewhart chart.

4. A MULTIVARIATE EXAMPLE

The general procedure that we have described above can easily be extended to the multivariate case. To illustrate the approach, consider a process manufacturing titanium castings, on which two vane openings (X and Y) are critical functional parameters. Since the vanes are on the same casting, it is reasonable to suspect that X and Y are related and that they should be controlled jointly. Measurements are made on each casting in time order of production with a coordinate measuring machine, and, consequently, serial correlation is suspected as well.

A sample of 50 measurements indicates that X and Y can be modeled by the multivariate ARMA process

$$X_t = 50 + 0.5X_{t-1} + 0.3Y_{t-1} + \varepsilon_t - 0.2\varepsilon_{t-1} - 0.1\eta_{t-1}$$

$$Y_t = 43 + 0.5y_{t-1} + 0.3X_{t-1} + \eta_t$$

where ε_t and η_t are white noise processes. Figures 8 and 9 present 100 measurements on X and Y, respectively. Both charts also display the Shewhart-type control limits for individual measurements, calculated from the original 50 observations. Both charts indicate out-of-control signals during both the initial 50 and final 50 observations.

Figures 10 and 11 present the univariate control charts for individual measurements for the residuals of X and Y, respectively. Now the general indication is that both variables are in control during the first 50 observations, and that the mean of X shifts downward following casting number 50. However, only one re-

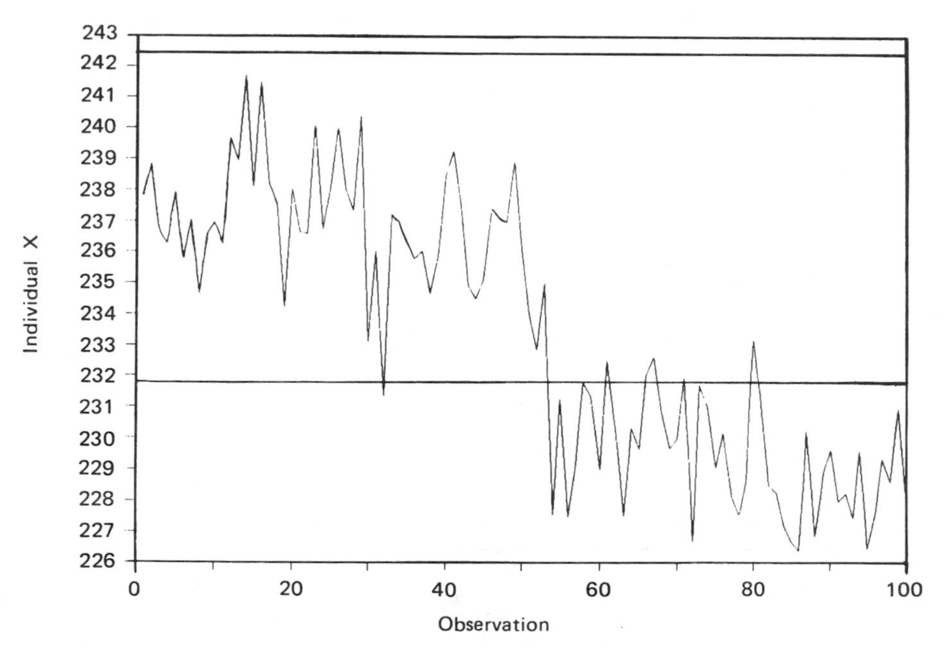

FIGURE 8 Control chart for X, original observations.

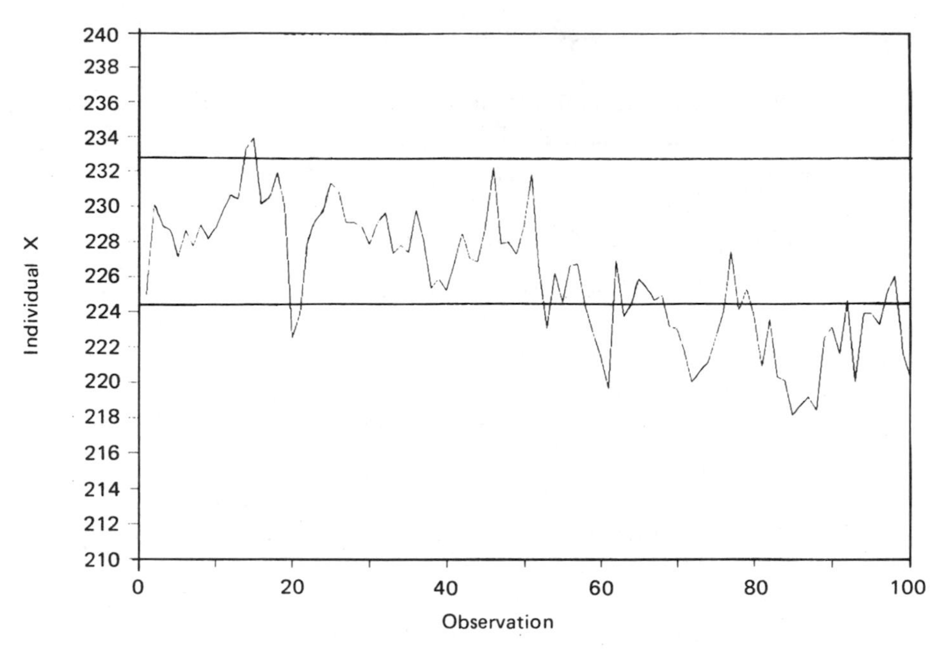

FIGURE 9 Control chart for *Y*, original observations.

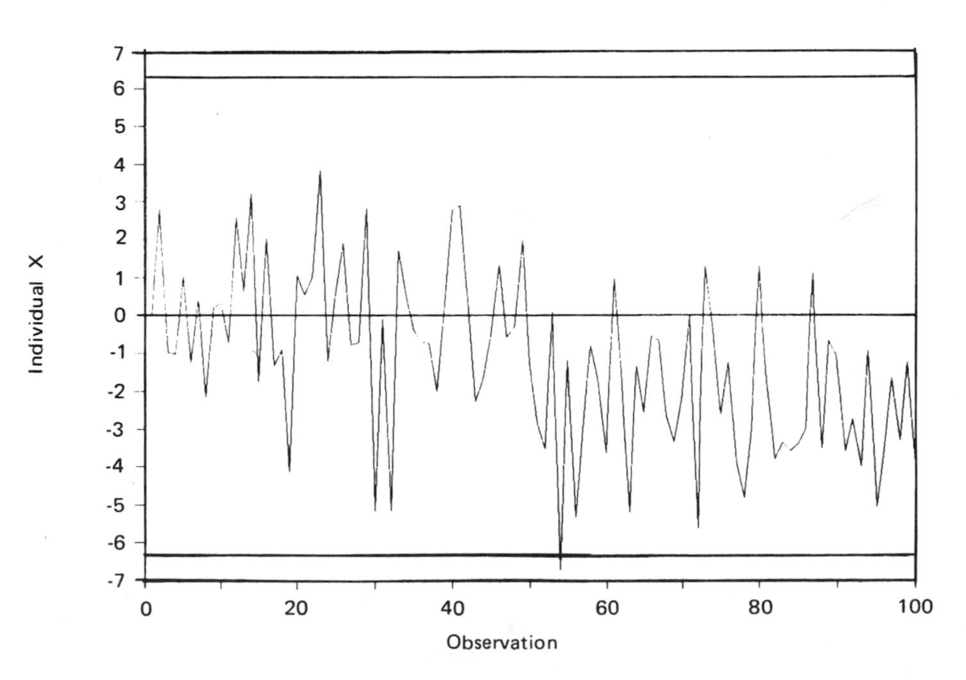

FIGURE 10 Control chart for *X*, residual values.

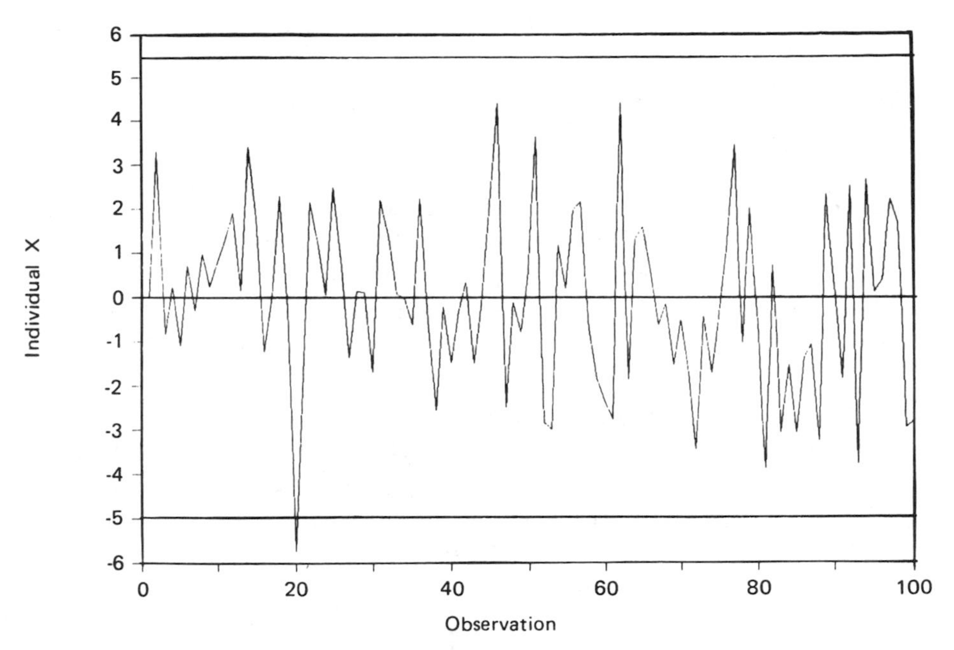

FIGURE 11 Control chart for *Y*, residual values.

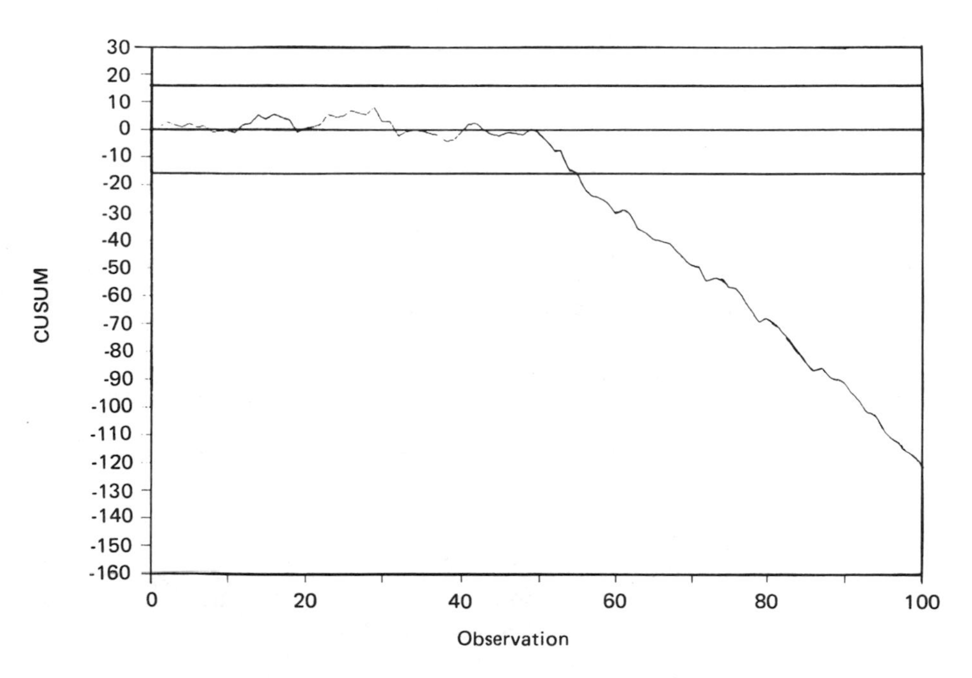

FIGURE 12 CUSUM for *X*, residual values.

sidual plots beyond the three-sigma control limit, so some other sensitizing rules may need to be used with this chart.

Figures 12 and 13 present the CUSUM control charts for the X and Y residuals, and Figures 14 and 15 present the GMA control charts for X and Y residuals, respectively. These charts clearly indicate that the process was in control until observation 50, and that the mean of X has shifted downward following casting 50. Once again, the GMA and CUSUM have provided a more rapid detection of the shift.

It is also possible to apply multivariate control charts directly to the residuals. In general, suppose that $\mathbf{e}' = [e_1, e_2, \ldots, e_p]$ is a vector of the residuals from a multivariate ARMA process, and that $\boldsymbol{\Sigma}$ is the covariance matrix of $\mathbf{e}$. Two types of control charts could be directly applied to $\mathbf{e}$. The first of these is the Hotelling T^2 chart. The statistic plotted on the chart is

$$T_t^2 = (\mathbf{e}_t - \bar{\mathbf{e}})'\boldsymbol{\Sigma}^{-1}(\mathbf{e}_t - \bar{\mathbf{e}}) \tag{10}$$

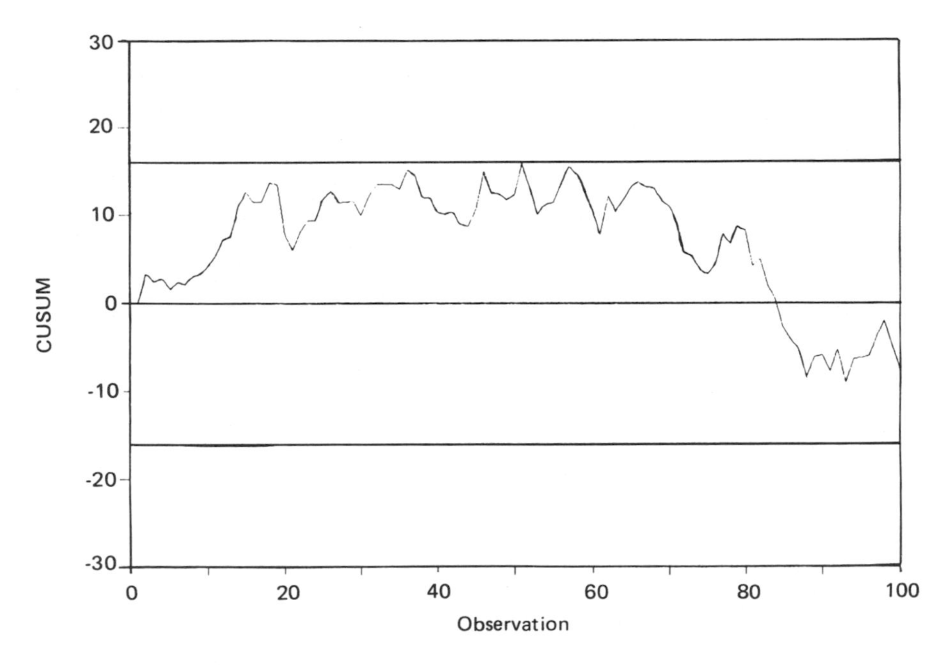

FIGURE 13 CUSUM for Y, residual values.

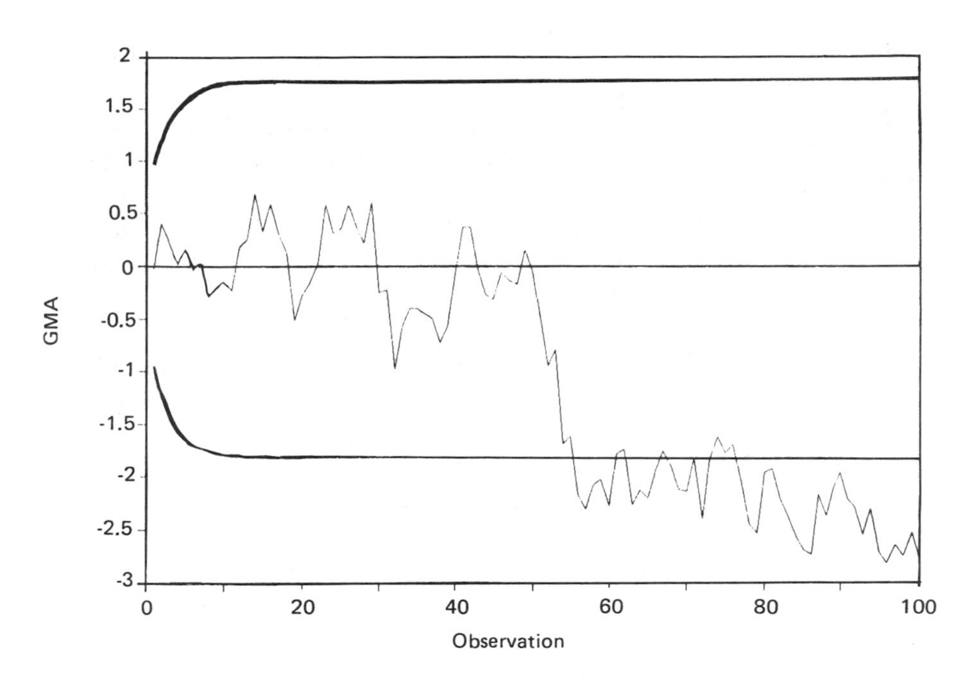

FIGURE 14 GMA for *X*, residual values.

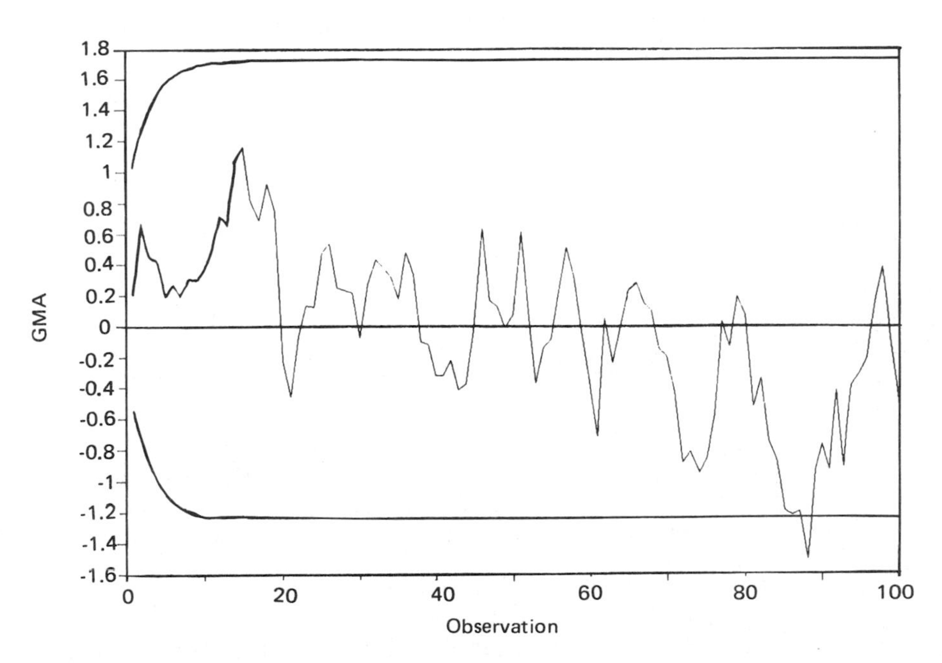

FIGURE 15 GMA for *Y*, residual values.

The Hotelling T^2 chart has only an upper control limit of $\mathrm{UCL} = \chi^2_{\alpha,p}$ assuming that $\boldsymbol{\Sigma}$ is known. For further details of this chart, see Montgomery (1985, Ch. 7). The second type of control chart would be to use either a CUSUM or GMA. There are many different ways to set up the CUSUM or GMA for multivariate data. One form of the multivariate CUSUM is

$$C_t = \max\left\{\frac{1}{t}[0, (\mathbf{e}_t - \bar{\mathbf{e}})'\boldsymbol{\Sigma}^{-1}(\mathbf{e}_t - \bar{\mathbf{e}}) - p + (t-1)C_{t-1}]\right\} \quad (11)$$

The approximate upper control limit for this CUSUM is $\chi^2_{\alpha,p}$.

Figure 16 presents the Hotelling T^2 control chart applied to the residuals. In constructing this chart, the sample covariance matrix was computed for the first 50 observations and used as an estimate of the process covariance matrix. The upper control limit on this chart was found using the chi-square distribution assuming that the process covariance matrix was known. Notice that the Hotelling T^2 control chart does not perform reliably in this particular example. There are indications that the process is out-of-control during the first 50 samples, and there is no strong indication of a process shift in the second 50 observations. It is possible that the out-of-control

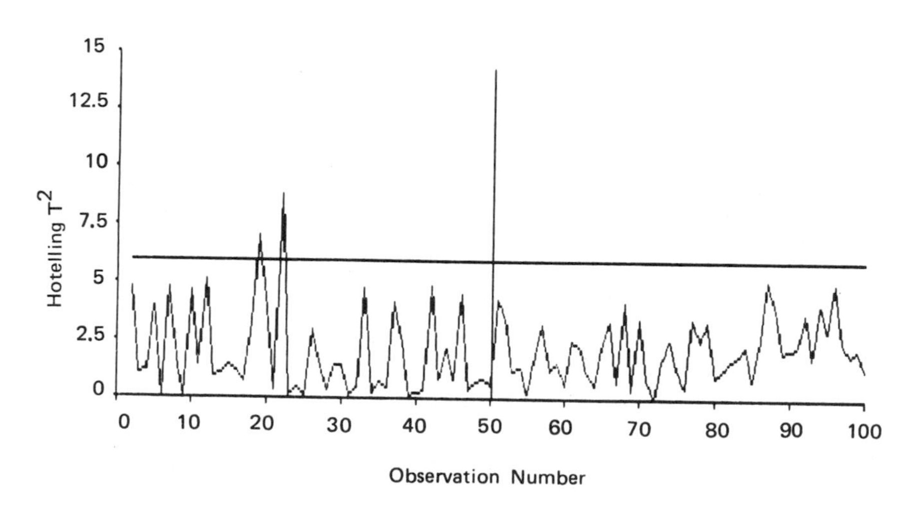

FIGURE 16 Multivariate data: residuals for the Hotelling T statistic.

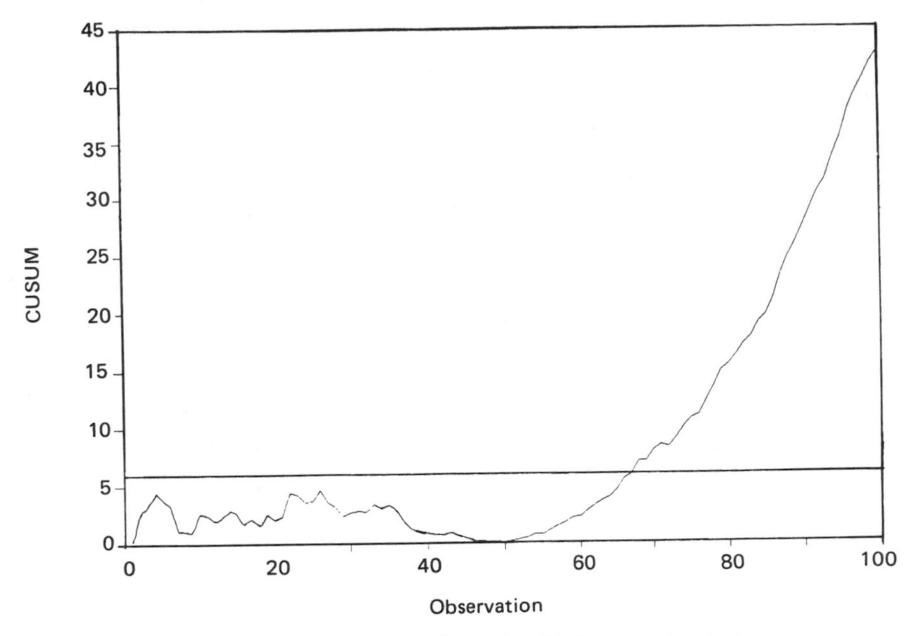

FIGURE 17 Multivariate CUSUM residual data.

signals during the first 50 observations arise because an inadequate estimate of the process covariance matrix was obtained.

The multivariate CUSUM chart is shown in Figure 17. Notice that this chart seems to perform extremely well. The chart indicates that the process is in control during the first 50 observations, and that a shift occurs following casting number 50. At this point, the analyst would have to investigate the individual measurements in order to determine exactly which characteristic has shifted out of control, and to determine when the shift had actually occurred.

5. CONCLUSION

This paper has illustrated how statistical process control methods can be applied in computer-integrated manufacturing, where serial correlation between sample measurements frequently occurs. The general approach indicated is to model the original signal with an appropriate empirical stochastic model, and then apply relatively standard control charts to the sequence of residuals generated from this model. We have illustrated how the procedure performs for both univariate and multivariate signals.

In practice, we have found that this approach works extremely well. The procedure is particularly effective in detecting shifts in location and dispersion in the original process. However, when the shift is of the form of a model shift [such as from AR(2) to AR(1)] and it is not accompanied by a corresponding change in either mean or variance, the procedures outlined in this paper are often inadequate. Control charts that directly examine the autocorrelative structure of the residuals would be required to detect this type of process deterioration. We have found that the GMA and CUSUM control charts are particularly effective for application to residuals. Since the GMA is somewhat more straightforward to interpret, and since it weights the signals so that older data receives less influence, it may be slightly preferable to the CUSUM. A well-designed GMA can give approximately the same average run length performance as the CUSUM. We have found that small values of r, typically around 0.1, provide excellent results. If a small value of r is used, then the width of the control limits on the GMA should be between 2.5 and 3 standard deviations. Since the GMA and CUSUM are most effective against small shifts, it is frequently desirable to accompany them with a control chart for individuals with moving range, so that large shifts may be detected more quickly.

REFERENCES

Bagshaw, M., and R. A. Johnson. (1975). The effect of serial correlation on the performance of CUSUM Tests II. *Technometrics, 17.*

Berthouex, T. M., W. G. Hunter, and L. Pallesen. (1978). Monitoring sewage treatment plants: Some quality control aspects. *Journal of Quality Technology, 10.*

Ermer, D. S. (1980). A control chart for dependent data. *ASQC 34th Annual Technical Conference Transactions*, Atlanta, Georgia.

Hunter, J. S. (1986). The exponentially weighted moving average. *Journal of Quality Technology, 18.*

Johnson, R. A., and M. Bagshaw. (1974). The effect of serial correlation on the performance of CUSUM tests. *Technometrics, 16.*

Liao, W. S., S. M. Wu, and D. S. Ermer. (1982). A time series approach to quality assurance. *Inspection and Quality Control in Manufacturing Systems.* Vol. 5. ASME.

Lucas, J. M. (1976). The design and use of cumulative sum quality control schemes. *Journal of Quality Technology, 8.*

Montgomery, D. C. (1985). *Introduction to Statistical Quality Control.* Wiley, New York.

Notohardjono, B. D., and D. S. Ermer. (1986). Time series control charts for correlated and contaminated data. *Journal of Engineering for Industry, 108*, 226.

Vasilopoulus, A. V., and A. P. Stamboulis. (1978). Modification of control chart limits in the presence of data correlation. *Journal of Quality Technology, 10.*

6
The Relationship Between Certain Kalman Filter Models and Exponential Smoothing Models

Nancy J. Kirkendall
U.S. Department of Energy
Washington, D.C.

1. INTRODUCTION

Exponential smoothing models (exponentially weighted moving average models) have been widely used to provide adaptive estimates of a mean function for management and forecasting, as well as for industrial applications. A recent paper by Hunter (1986) describes the use of control charts for the manufacturing industries based on exponential smoothing and shows their relationship to Shewhart and cumulative sum (CUSUM) charts. Sweet (1986) provides a number of references to the use of exponential smoothing in the construction of control charts and discusses coupled exponential smoothing equations to monitor both the mean and the variance of a process.

Exponential smoothing has been used extensively as a model of a mean function because it provides a simple summary of past data and has a simple update equation. Thus, it minimizes both computer data storage requirements and computation time. Additionally, it has been found to perform well in a wide variety of situations, even though the parameter selection has generally been left to the analyst's judgement.

The Kalman filter recursion equations also provide a simple summary of past data, and the form of the update equation is relatively simple. However, the Kalman filter is much more powerful. The Kalman filter recursion equations can be used to implement exponential smoothing models as will be described in this paper. They can also be used to implement much more complicated models, as described by Takata et al. (1985), in which a Kalman filter was used to implement an autoregressive model with 27 parameters to monitor for tool breakage.

This paper discusses two exponential smoothing models, simple exponential smoothing and exponential smoothing with trend. For each model, it describes the autoregressive integrated moving average (ARIMA) model that is equivalent to it and the Kalman filter model that is asymptotically equivalent. Most importantly, the relationships between the parameters in the various forms of the models are also shown. This information should assist users in selecting the variances that are required inputs to the Kalman filter models.

There are at least two advantages to a Kalman filter implementation of an exponential smoothing model. First, it can be started with no data. Second, if an intervention results in a change to the mean at some known point in time, the Kalman filter can be made to adapt immediately by a simple change to the variance of the prior state estimate at that time.

The relationships between exponential smoothing models, ARIMA models, and the related Kalman filter models have been discussed by a number of authors over the years. A 1967 paper by Harrison was one of the first. That paper addressed simple forecasting procedures based on trend models. In their paper on bayesian forecasting, Harrison and Stevens (1976) introduce the two Kalman filter models for the trend that will be discussed in this paper. (Harrison and Stevens refer to the Kalman filter models as "dynamic linear models.") The two models discussed in this paper are the "steady model," asymptomatically equivalent to simple exponential smoothing and to an ARIMA(0, 1, 1) model; and the "linear growth" model, asymptomatically equivalent to Holt's exponential smoothing with trend and to an ARIMA(0, 2, 2) model. A more recent paper (1986) by Abraham and Ledolter describes these relationships more thoroughly, and includes a discussion of models that include both trend and seasonal components.

In recent years, trend models such as the "steady" model and the "linear growth" model have also been incorporated in structural models for time series. Examples are Durbin (1986), Maravall (1985), Harvey and Todd (1983), and Gersch and Kitagawa (1983).

This paper outlines one of the standard derivations of the Kalman filter recursion equations in Section 1. It illustrates the model relationships for the simple exponential smoothing model (the "steady" model) in Section 2, and describes the relationships for the exponential smoothing with trend model (the "linear growth" model) in Section 3. The new information concerns identifying the variance inputs required for the Kalman filter models based on the desired smoothing coefficients, and an illustration of the rate of convergence of the smoothing coefficient of the Kalman filter. Additionally, the alternative "linear growth" model used by Gersch and Kitigawa (1983) is considered. Its relationship to the "linear growth" model used by other authors, to exponential smoothing with trend, and to an ARIMA(0, 2, 2) model is shown in Section 4.

1.1 The Kalman Filter

The Kalman filter recursion equations can be derived in a number of ways. One way, outlined below, is a Bayesian procedure that makes use of the assumption that the errors are normally distributed. The same equations can be derived without the assumption of normality by solving for the minimum variance, unbiased linear estimator.

1.2 The State and Measurement Equations

The following notation for the Kalman filter equations is taken from Gelb (1979).

Let the state equation be defined as

$$\mathbf{x}_t = \phi(t-1)\mathbf{x}_{t-1} + \mathbf{w}_{t-1} \tag{1}$$

where $\mathbf{x}_t$ is the $p \times 1$ state vector at time t, $\phi(t)$ is the $p \times p$ transition matrix at time t, which is assumed to be known for all t, and $\mathbf{w}_t$ is the $p \times 1$ error vector. The error vector is assumed to be normally distributed with mean vector $\mathbf{0}$ and $p \times p$ variance matrix $Q(t)$, which is assumed to be known for all t. At any point in time, $\mathbf{x}_t$ follows a normal distribution independent of $\mathbf{w}_t$.

The observation equation is

$$\mathbf{y}_t = H(t)\mathbf{x}_t + \mathbf{v}_t \tag{2}$$

where $\mathbf{y}_t$ is the $n \times 1$ vector of observations at time t, $H(t)$ is the $n \times p$ measurement matrix, which is assumed to be known, and $\mathbf{v}_t$ is the $n \times 1$ error vector. The error vector is also assumed to be normally distributed with mean vector $\mathbf{0}$ and $n \times n$ variance matrix $R(t)$, which is also assumed to be known for all t. The error vectors $\mathbf{w}_t$ and $\mathbf{v}_k$ are assumed to be uncorrelated for all t and k.

1.3 Initiating the Recursion

One advantage of the Kalman filter recursion relationship is that the observations $\mathbf{y}_t$ are processed one at a time. The procedure operates by first forecasting to get a prior estimate of the state at some point in time, and then updating that forecast with the most recent observation.

The initial estimate for the characteristics of the state at time $t = 0$ must be supplied by the user in order to provide a starting point. To initiate the process, it is assumed that the mean and variance of $\mathbf{x}_0$ are given by the fixed values $\hat{\mathbf{x}}_0$ and P_0. These values and the state equation (1) provide a prior distribution for the state at time $t = 1$. The distribution is normal with mean

$$\mathbf{x}_1(-) = \phi(0)\hat{\mathbf{x}}_0, \tag{3}$$

and variance

$$P_1(-) = \phi(0)P_0\phi(0)' + Q(0) \tag{4}$$

where the prime denotes transpose. The notation $(-)$ indicates that these parameters are associated with a prior estimate for the parameters of the state.

In general, given the mean and variance of the distribution of $\mathbf{x}_{t-1}$ that makes use of the data up to time $t - 1$, say, $\mathbf{x}_{t-1}(+)$ and $P_{t-1}(+)$, the same forecasting procedure is used to obtain a prior estimate for the mean and variance of the state at time t. For general t, the equations are

$$\mathbf{x}_t(-) = \phi(t-1)\mathbf{x}_{t-1}(+) \tag{5}$$

and

$$P_t(-) = \phi(t-1)P_{t-1}(+)\phi(t-1)' + Q(t-1) \tag{6}$$

1.4 Updating the Prior Estimate with Data

Equation (2) implies that the conditional distribution of $\mathbf{y}_t|\mathbf{x}_t$ is normal, with mean $H(t)\mathbf{x}_t$ and variance $R(t)$. Given the prior distribution of $\mathbf{x}_t$ created by forecasting, and the conditional distribution of $\mathbf{y}_t|\mathbf{x}_t$, the posterior distribution of the state at time t has mean

$$\mathbf{x}_t(+) = [I - K(t)H(t)]\mathbf{x}_t(-) + K(t)\mathbf{y}_t \tag{7}$$

and variance

$$P_t(+) = [I - K(t)H(t)]P_t(-) \tag{8}$$

The notation $(+)$ indicates that the parameters are associated with the posterior distribution of the state, given the data up to time t. The matrix $K(t)$ in Eqs. (7) and (8) is evaluated as

$$K(t) = P_t(-)H(t)'[H(t)P_t(-)H(t)' + R(t)]^{-1} \tag{9}$$

or equivalently as

$$K(t) = P_t(+)H(t)'R(t)^{-1}, \tag{10}$$

where the superscript -1 denotes matrix inversion. With this definition of $K(t)$, Eqs. (7) and (8) yield the mean and the variance of $\mathbf{x}_t|\mathbf{y}_t$. The term $K(t)$ is called the Kalman gain matrix.

2. SIMPLE EXPONENTIAL SMOOTHING: THE STEADY MODEL

In a simple exponential smoothing model the mean, $x_t(+)$, is calculated as

$$x_t(+) = (1 - \alpha)x_{t-1}(+) + (\alpha)y_t \tag{11}$$

where alpha is the smoothing coefficient and has some value between zero and one, and y_t is the observation at time t.

This is equivalent to an ARIMA(0, 1,1) model with parameter theta $= (1 - \alpha)$, as long as theta is greater than zero.

The Kalman filter model that is equivalent to the simple exponential smoothing model has state equation

$$x_t = x_{t-1} + w_{t-1} \tag{12}$$

where w_t is $N(0, q)$. The state equation models the behavior of the mean of the series. The measurement equation is

$$y_t = x_t + v_t \tag{13}$$

where v_t is $N(0, r)$. The measurement equation relates the observation, y_t, to the current mean of the series, x_t.

Both of these equations are univariate, so that in Eqs. (1) and (2) $p = 1$ and $n = 1$. Further, $\phi(t) = 1$, $H(t) = 1$, $Q(t) = q$, and $R(t) = r$ for all t. With this system of equations, the forecast equation (5) can be substituted into the update equation (7) to get the simple univariate expression

$$x_t(+) = [1 - K(t)]x_{t-1}(+) + K(t)y_t \tag{14}$$

By comparing the coefficients in Eqs. (14) and (11) it is easy to see that this Kalman filter model is equivalent to an exponential smoothing model, with time-varying parameter $K(t)$, the Kalman gain from Eqs. (9) or (10). For this model Eq. (10) reduces to

$$K(t) = P_t(+)/r \tag{15}$$

and substituting Eq. (6) into Eq. (9) yields

$$K(t) = [P_{t-1}(+) + q]/[P_{t-1}(+) + q + r] \tag{16}$$

The steady-state value of the posterior variance $P(+)$ is found by setting $P_t(+) = P_{t-1}(+) = P(+)$ in Eqs. (15) and (16). Setting

Eq. (15) equal to Eq. (16) and solving for $P(+)$ yields

$$P(+) = [-1 + \text{sqrt}(1 + 4r/q)]q/2 \tag{17}$$

The associated steady-state Kalman gain, $K = P(+)/r$, is

$$K = [-1 + \text{sqrt}\,(1 + 4r/q)]q/(2r) \tag{18}$$

The terms $P_t(+)$ and $K(t)$ are convergent monotone functions in t. Thus, if P_0 is smaller than $P(+)$ then $K(t)$ will be a monotone increasing bounded function, and will converge from below to Eq. (18). If P_0 is larger than $P(+)$ then $K(t)$ will be a monotone decreasing bounded function and will converge from above to Eq. (18). If $P_0 = P(+)$, then the Kalman filter results are identical to the simple exponential smoothing model, with smoothing coefficient alpha equal to K.

The asymptomatic variance, Eq. (17), depends only on the value assigned to q and on the ratio r/q. The asymptomatic smoothing coefficient K depends only on the ratio r/q. Thus, the "steady" Kalman filter model is asymptomatically a simple exponential smoothing model with asymptomatic coefficient given by Eq. (18). Note that if $q = 0$, then the asymptomatic smoothing coefficient is zero. In this case, this Kalman filter model asymptomatically yields the mean of the data as the state estimate. If P_0 is set to infinity, $x_t(+)$ is the mean of the data for all t, and $P_t(+) = r/t$.

Harrison (1967) derived expression (18) as the value for the optimal exponential smoothing coefficient in the steady model (the coefficient that minimizes the one-step forecast error). This is consistent with the property that the Kalman filter yields the minimum variance estimator. The equivalence between the "steady model," an ARIMA(0, 1, 1) model, and simple exponential smoothing is also demonstrated in Abraham and Ledolter (1986).

2.1 Practical Implications

One way of evaluating the adequacy of this Kalman filter model is to fit an ARIMA(0, 1, 1) model to the data and use the standard diagnostics. If the model is adequate, and theta is positive, then the value of q/r that yields the asymptomatic smoothing coefficient K equal to $(1 - \theta)$ is optimal.

Given the optimal smoothing coefficient, a plot of Eq. (18) as a function of q/r can be used to select the variance ratio that leads to the desired smoothing coefficient (see Figure 1). Smoothing coefficients 0.1 and 0.5, for example, are related to variance ratios of $q/r = 1/90$ and $q/r = 1/2$, respectively.

Figures 2 and 3 illustrate the rates of convergence of $K(t)$ to these steady-state values for a variety of values of the starting variance P_0. Note that for P_0 infinite (or very large) the initial value $K(1) = 1$, so that the first observation receives weight 1, and the prior value receives weight 0. This is appropriate unless prior information actually provides a reasonable estimate for the initial value of the state. Setting $P_0 = rK$ yields the smoothing coefficient K for all t. In this case, the prior mean of the state x_0 is assumed to be a good estimate for the mean of the distribution of the state at time $t = 0$.

Note that for a wide range of values of P_0, convergence to the steady-state value is quite fast, and is faster for the larger values of K. Figure 2 shows that for $K = 0.1$, $K(t)$ is approximately equal to K by $t = 15$. Figure 3 shows that for $K = 0.5$, $K(t)$ is approximately equal to K by $t = 5$.

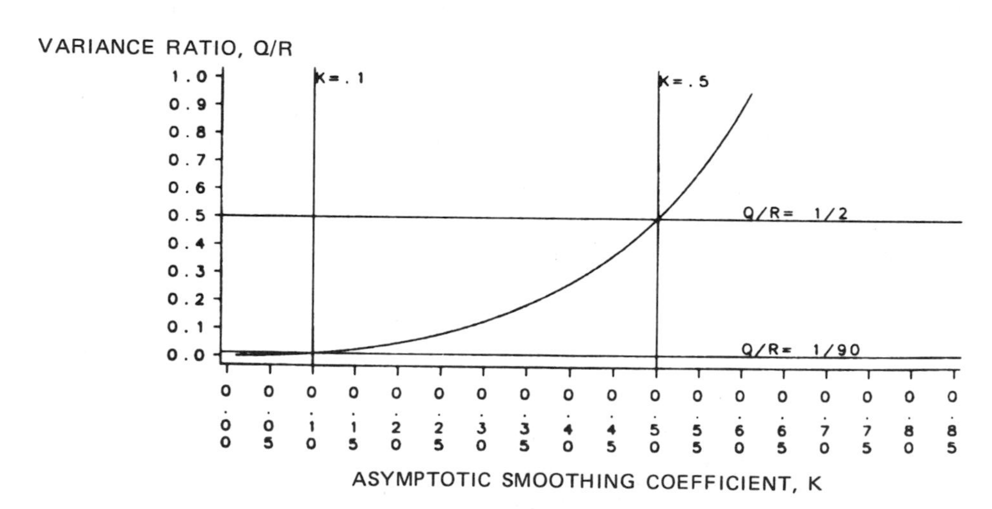

FIGURE 1 Relationship between Kalman gain and variance ratio for the steady model.

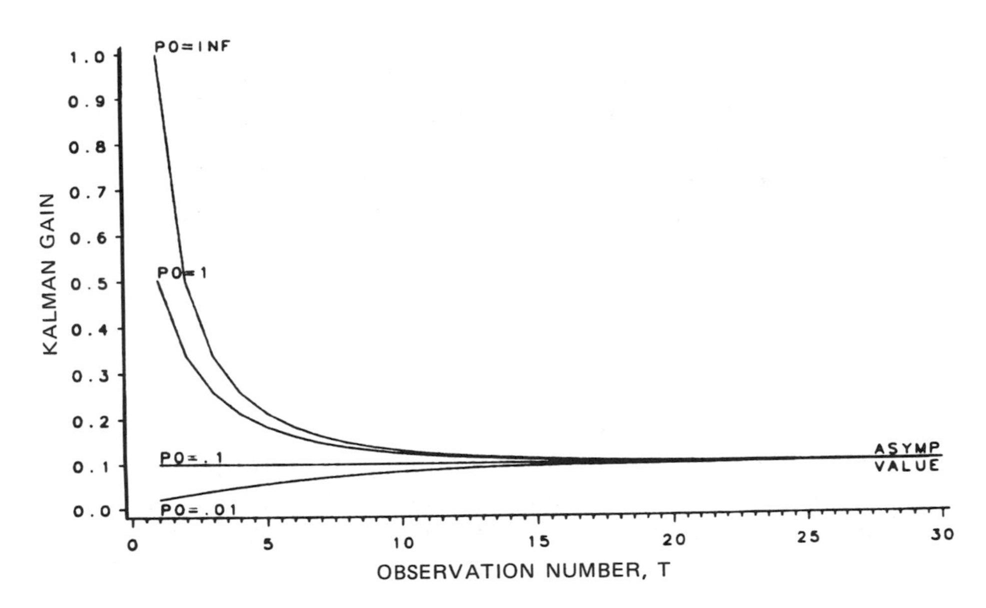

FIGURE 2 Kalman gain for steady model, asymptotic smoothing coefficient = 0.1 ($r = 1$, $q = 0.011111$).

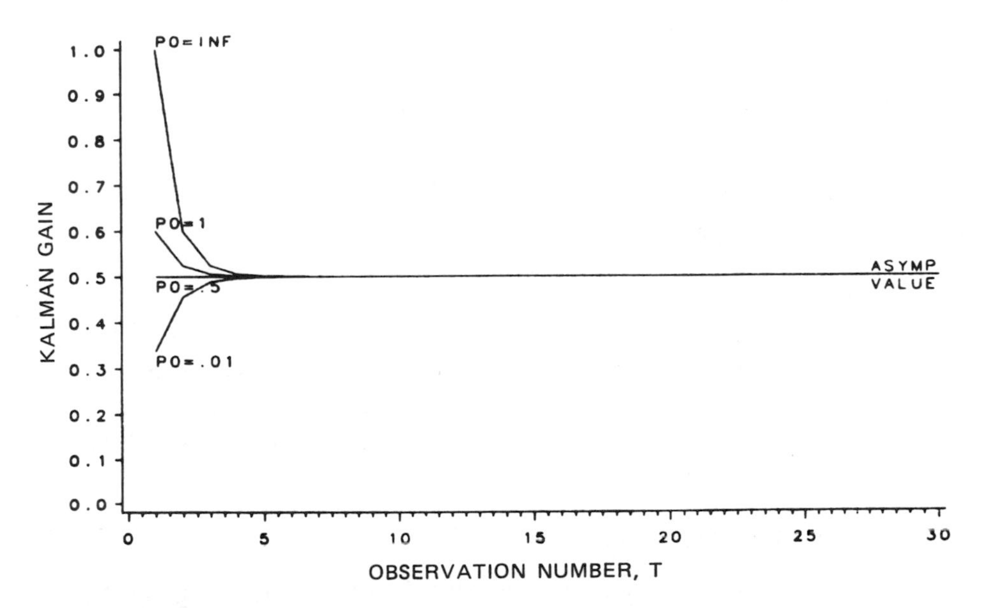

FIGURE 3 Kalman gain for steady model, asymptotic smoothing coefficient = 0.5 ($r = 1$, $q = 0.5$).

Harrison and Stevens (1976) present two advantages of the Kalman filter implementation of the simple exponential smoothing model. First, as mentioned above, it can be started with no data; second, it can be used to model an intervention. They present the following as an example of the later. In a sales forecasting system with a major change in the market at time t_0, the new unknown mean of the series could be tracked by changing the variance of the prior distribution at that time. If the prior variance at time t_0, $P_{t_0}(-)$, was changed to some very large number, the exponential smoothing coefficient would give a weight close to one to the observation at t_0, and almost no weight to the previous mean. From the value $K(t_0) = 1$, the smoothing coefficient would converge back to the steady-state value. In this way, the estimate for the state would converge rapidly to the new postintervention value.

3. EXPONENTIAL SMOOTHING WITH TREND: THE LINEAR GROWTH MODEL

Exponential smoothing with trend uses two coefficients and estimates two terms: $a_t(+)$, the current level of the state, and $b_t(+)$, the growth term. (The growth represents the expected increase or decrease between adjacent observations.) The exponential smoothing equations are

$$a_t(+) = Ay_t + (1 - A)[a_{t-1}(+) + b_{t-1}(+)]$$

$$b_t(+) = C[y_t - a_{t-1}(+)] + (1 - C)b_{t-1}(+) \tag{19}$$

The smoothing coefficients are A and C, and y_t is the observation at time t. The coefficients A and C must be in the interval from zero to one.

This is equivalent to an ARIMA(0, 2, 2) model with parameters $A = (-\theta_2)$, and $C = (1 - \theta_1 - \theta_2)$, when theta$_2$ is negative and $(\theta_1 + \theta_2)$ is positive. This equivalence is demonstrated in Abraham and Ledolter (1986).

3.1 The Standard Formulation

The Kalman filter state equation related to this model is given by setting

$$\mathbf{x}_t = \begin{bmatrix} a_t \\ b_t \end{bmatrix} \qquad \phi(t) = \begin{bmatrix} 1 & 1 \\ 0 & 1 \end{bmatrix} \qquad \mathbf{w}_t = \begin{bmatrix} w_1(t) \\ w_2(t) \end{bmatrix}$$

in Eq. (1), yielding the bivariate state equation

$$\begin{bmatrix} a_t \\ b_t \end{bmatrix} = \begin{bmatrix} 1 & 1 \\ 0 & 1 \end{bmatrix} \begin{bmatrix} a_{t-1} \\ b_{t-1} \end{bmatrix} + \begin{bmatrix} w_1(t) \\ w_2(t) \end{bmatrix} \tag{20}$$

where the error term $\mathbf{w}_t$ is bivariate $N(0, Q)$, and

$$Q = \begin{bmatrix} q_{11} & q_{12} \\ q_{12} & q_{22} \end{bmatrix}$$

The univariate measurement equation is given by setting $H(t) = (1, 0)$ in Eq. (2), which yields

$$y_t = (1, 0) \begin{bmatrix} a_t \\ b_t \end{bmatrix} + v_t \tag{21}$$

where v_t is normal $(0, r)$.

The 2×2 covariance matrix of the prior distribution of the state at time $t = 0$ is denoted P_0.

The Kalman recursive equations yield the same model as the exponential smoothing with trend, but with time-varying coefficients. These are

$$A_t = p_{11t}(+)/r \qquad C_t = p_{21t}(+)/r \tag{22}$$

where $p_{11t}(+)$ denotes the (1, 1) element of $P_t(+)$ in Eq. (8), and $p_{21t}(+)$ denotes the (2, 1) element of the same matrix.

The steady-state solution for the 2×2 matrix $P(+)$, where $P(+) = P_t(+) = P_{t-1}(+)$ for sufficiently large t, can be found by solving the equation

$$Q = [P(+)^{-1} - H'R^{-1}H]^{-1} - \phi P(+)\phi' \tag{23}$$

for $P(+)$, or equivalently, by solving the system of equations

$$\begin{aligned} q_{11} &= p_{11}(+)^2/[r - p_{11}(+)] - [2p_{12}(+) + p_{22}(+)] \\ q_{12} &= p_{11}(+)p_{12}(+)/[r - p_{11}(+)] - p_{22}(+) \\ q_{22} &= p_{12}(+)^2/[r - p_{11}(+)] \end{aligned} \tag{24}$$

for $p_{11}(+)$, $p_{12}(+)$, and $p_{22}(+)$, the elements of $P(+)$.

Note that the third equation in Eq. (24) yields an expression for $p_{12}(+)$ in terms of $p_{11}(+)$, and the second equation yields an expression for $p_{22}(+)$ in terms of $p_{11}(+)$ and $p_{12}(+)$. These are

$$p_{12}(+) = \text{sqrt}\,[q_{22}(r - p_{11}(+)]$$

and

$$p_{22}(+) = p_{11}(+)p_{12}(+)/[r - p_{11}(+)] - q_{12}$$

Substituting into the first equation in Eq. (24) yields the solution of $p_{11}(+)$ to be one of the roots of the quartic equation

$$z^4 + z^3(2b + a) + z^2[b^2 - 2r(2b + 5a)] + z(8ar^2 - 2rb^2) + r^2b^2 - 4ar^3 = 0 \quad (25)$$

where $a = q_{22}$ and $b = (q_{11} - q_{12})$.

These steady-state values for the matrix $P(+)$ yield the steady-state smoothing coefficients

$$A(+) = p_{11}(+)/r$$

$$C(+) = p_{12}(+)/r = \text{sqrt}\,\{[(1 - A(+)]q_{22}/r\} \quad (26)$$

Equation (25) can be rewritten in terms of $A(+)$. The steady-state solution for the smoothing coefficient $A(+)$ is one of the roots of

$$A(+)^4 + A(+)^3(2d + c) + A(+)^2(d^2 - 2d - 5c) + A(+)(-2d^2 + 8c) + d^2 - 4c = 0 \quad (27)$$

where $c = q_{22}/r$ and $d = (q_{11} - q_{12})/r$.

Note that the solution for $A(+)$ depends only on the ratio of elements of Q to r. Additionally, values of q_{11} and q_{12} enter only as the difference $q_{11} - q_{12}$.

The system of equations in Eq. (24) is equivalent to the system describing the optimal solution in Harrison (1967), when $q_{12} = 0$.

3.2 Practical Implications

Model adequacy can be evaluated by using the diagnostics associated with fitting an ARIMA(0, 2, 2) model to the data. If the model is adequate, theta_2 is negative, and $(\text{theta}_1 + \text{theta}_2)$ is positive, then the optimal smoothing coefficients would be given by $A(+) = -\text{theta}_2$ and $C(+) = 1 - (\text{theta}_1 + \text{theta}_2)$. Given these values, the variance ratios can be selected from a plot such as the one in Figure 4.

Figure 4 fixes $A(+)$ in Eq. (27) to the value of the desired smoothing coefficient, $A(+) = 0.5$ in this example. Note that for a

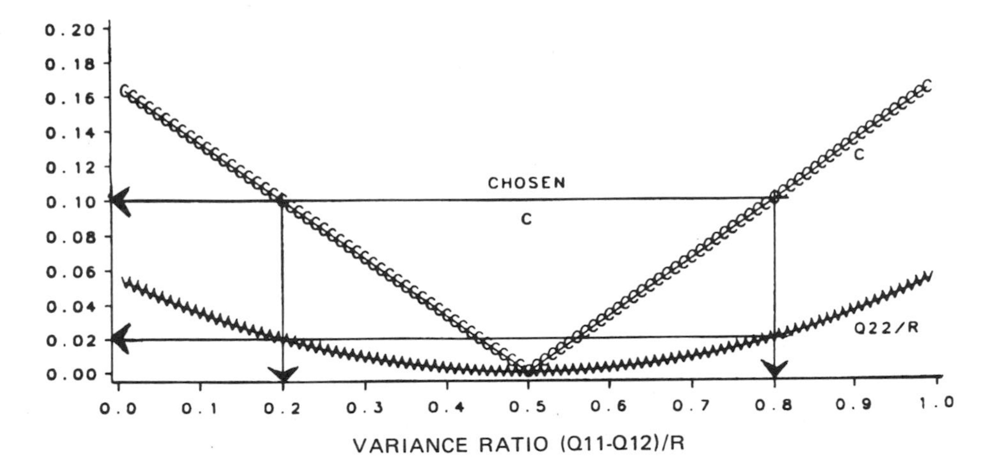

FIGURE 4 Nomogram for selecting variance ratios to yield desired smoothing coefficients for linear growth model; C is smoothing coefficient for trend, A is smoothing coefficient for level, and $A = 0.5$.

given value of $A(+)$ and a given value of $d = (q_{11} - q_{12})/r$, Eq. (27) is a linear function of the remaining variance ratio, q_{22}/r. Figure 4 is calculated by fixing $A(+)$ to the desired level, iterating through values of d from 0.01 to 1 in increments of 0.01, and solving Eq. (27) for the value of c. Additionally, given the values of $A(+)$, c, and d, Eq. (26) gives the value of the smoothing coefficient for the trend, $C(+)$. If the desired smoothing coefficient for the trend is $C(+) = 0.1$, then this chart shows that there are two sets of variance ratios that give the desired smoothing coefficients: both have $q_{22}/r = 0.02$, but one has $(q_{11} - q_{12})/r = 0.2$, and the other has variance ratio $(q_{11} - q_{12})/r = 0.8$.

Figures 5 and 6 illustrate the rates of convergence to the pure exponential smoothing models for a variety of values of P_0, for the first set of variance ratios. Note that convergence is still quite fast, with the coefficients approximately equal to the steady-state values by $t = 10$. Note also that to obtain the steady-state values for all values of t, P_0 must be initialized to $P(+)$, which is not generally a diagonal matrix. (This was done in the $P_0 = 0.5$ case.)

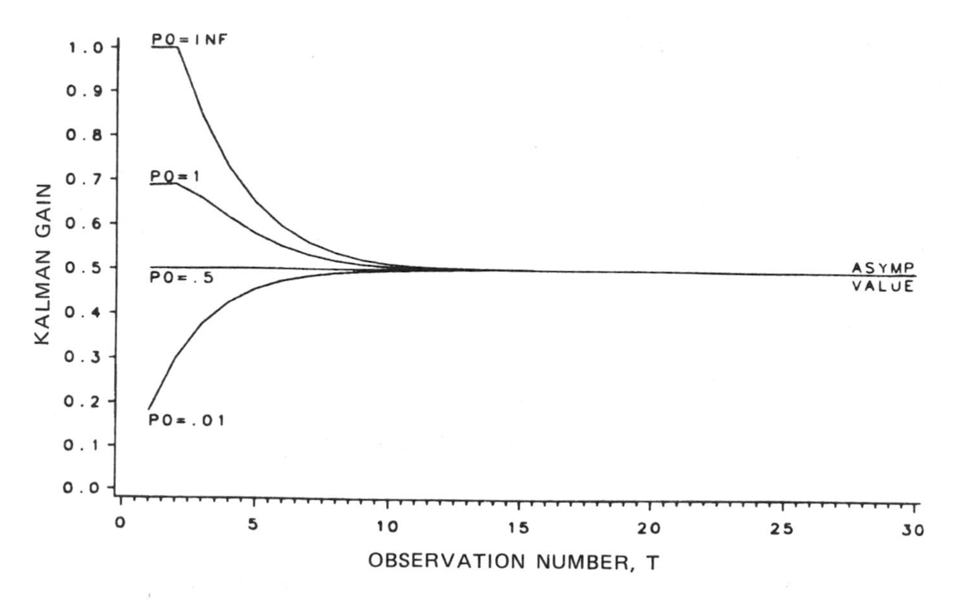

FIGURE 5 Kalman gain for linear growth model, asymptotic smoothing coefficient = 0.5 ($r = 1$, $q_{11} = 0.2$, $q_{22} = 0.02$).

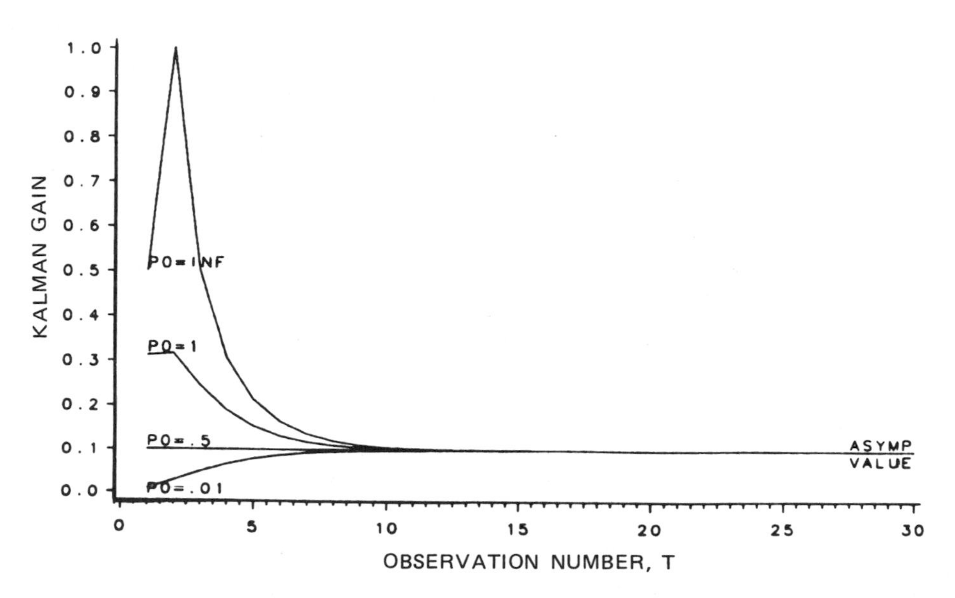

FIGURE 6 Kalman gain for linear growth model, asymptotic smoothing coefficient for trend = 0.1 ($r = 1$, $q_{11} = 0.2$, $q_{22} = 0.02$).

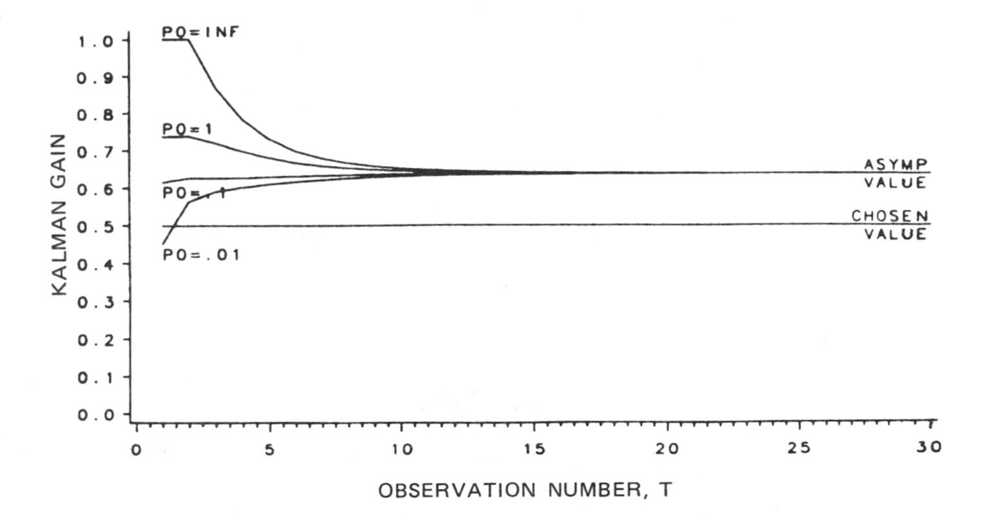

FIGURE 7 Kalman gain for linear growth model, asymptotic coefficient for level = 0.5 ($r = 1$, $q_{11} = 0.8$, $q_{22} = 0.02$), example of case when quartic equation has more than one root in (0, 1) for given variance ratios.

Figure 7 illustrates the convergence of the parameter $A_t(+)$ for a variety of starting values P_0 for the second set of variance ratios identified in Figure 4. Note here that although $A_t(+)$ clearly converges, it does not converge to $A(+) = 0.5$, as desired. This illustrates the fact that $A_t(+)$ will converge to one of the roots of a quartic equation. If there is only one root in the interval (0, 1) then there is no problem identifying the required variance ratios. If there is more than one root in the interval (0, 1) then it appears that $A_t(+)$ converges to the largest such root.

4. AN ALTERNATIVE FORMULATION FOR THE LINEAR GROWTH MODEL

An alternative formulation for the Kalman filter equations representing exponential smoothing with trend is used by Gersch and Kitagawa (1983). The state equation for their system is given by setting

$$\mathbf{x}_t = \begin{bmatrix} x_t \\ x_{t-1} \end{bmatrix} \qquad \phi(t) = \begin{bmatrix} 2 & -1 \\ 1 & 0 \end{bmatrix} \qquad \mathbf{w}_t = \begin{bmatrix} w_{1t} \\ 0 \end{bmatrix}$$

in Eq. (1), yielding

$$\begin{bmatrix} x_t \\ x_{t-1} \end{bmatrix} = \begin{bmatrix} 2 & -1 \\ 1 & 0 \end{bmatrix} \begin{bmatrix} x_{t-1} \\ x_{t-2} \end{bmatrix} + \begin{bmatrix} w_{1t} \\ 0 \end{bmatrix} \tag{28}$$

where w_{1t} is $N(0, q)$. An alternative way of writing this is

$$(1 - B)^2 x_t = w_{1t}$$

where B is the backshift operator.

This model is also asymptomatically equivalent to the exponential smoothing with trend model, but in this case $A(+)$ is equal to $C(+)$. The same smoothing coefficient is applied to both trend and level estimates.

The steady-state value of this coefficient $A(+)$ will satisfy

$$A(+)^4(r/q) + A(+)^3 - 5A(+)^2 + 8A(+) - 4 = 0 \tag{29}$$

Note that for a given value of $A(+)$, there is only one variance ratio that solves Eq. (29). Figure 8 plots $A(+)$ along the horizontal

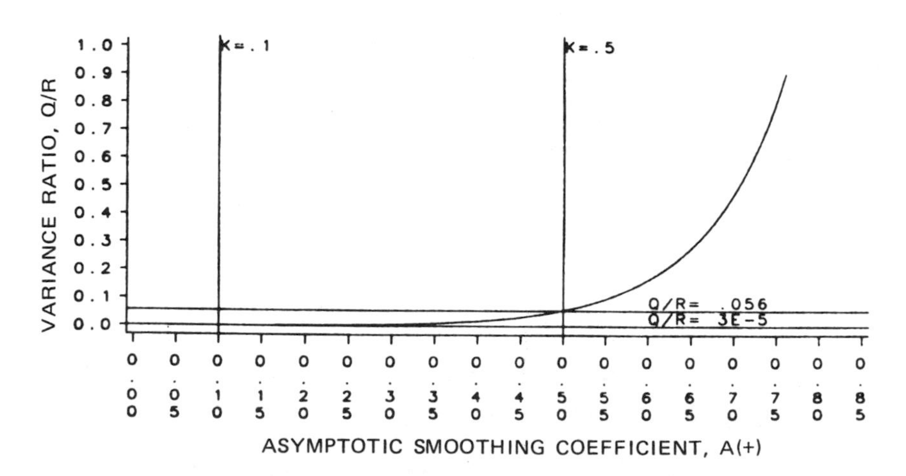

FIGURE 8 Relationship between Kalman gain and variance ratio for the linear growth model, Gersch–Kitagawa variation.

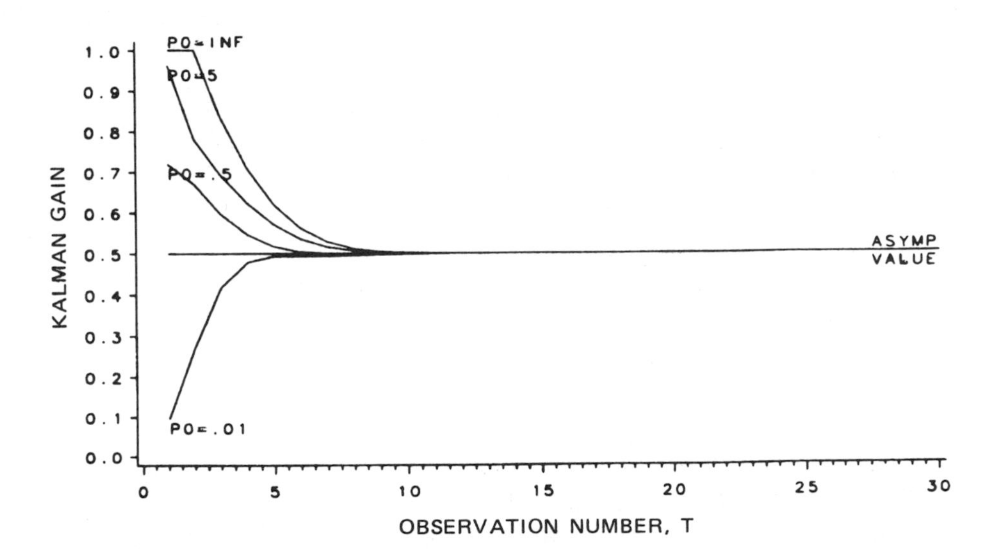

FIGURE 9 Kalman gain for linear growth model, Gersch–Kitagawa variation, asymptotic smoothing coefficient = 0.5 ($r = 1$, $q = 0.05555$).

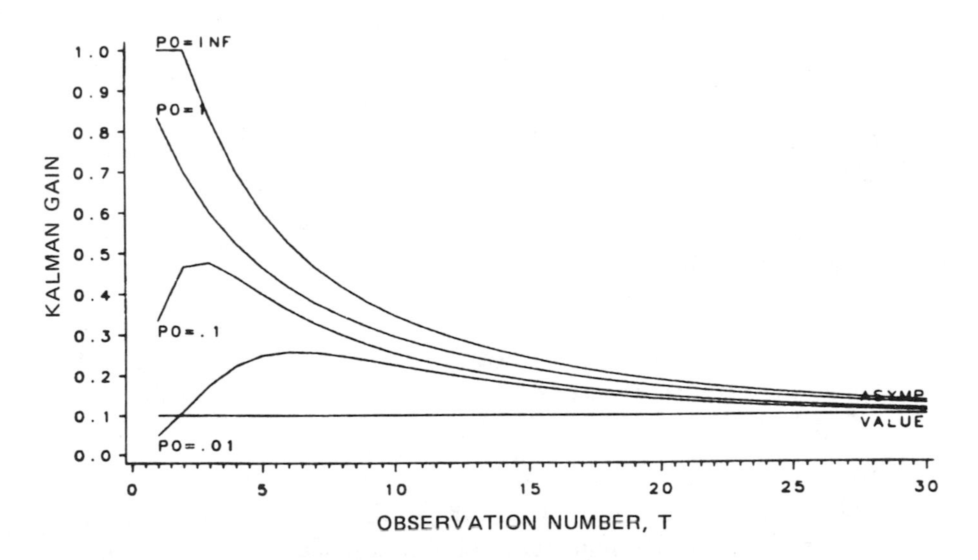

FIGURE 10 Kalman gain for linear growth model, Gersch–Kitagawa variation, asymptotic smoothing coefficient = 0.1 ($r = 1$, $q = 0.0000307$).

axis, and the variance ratio q/r along the vertical. This chart shows that for a smoothing coefficient of $A(+) = 0.5$, the variance ratio should be $q/r = 0.05555$, and for a smoothing coefficient of 0.1, the variance ratio should be $q/r = 0.0000307$. Convergence to these values is illustrated in Figures 9 and 10. Convergence to the steady-state value occurs by about $t = 10$ for $A(+) = 0.5$, and by about $t = 40$ for $A(+) = 0.1$.

5. CLOSING REMARKS

The Kalman filter models are beginning to receive more attention from the statistical community. The intent of this paper has been to promote this interest by showing the relationship of widely understood exponential smoothing models to certain Kalman filter models. When one knows the applicability of exponential smoothing models to statistical process control, it naturally follows that Kalman filter models might also be used. The benefits of these models are the minimum data requirements for start-up and the sensitivity to changes in the underlying conditions. Further research in applying these models may exploit these characteristics for monitoring small batch production or process subject to slight changes in location. Advantage may also be taken of the model's predictive capability to correct for process problems prior to actual observations.

REFERENCES

Abraham, B., and J. Ledolter. (1986). Forecast functions implied by autoregressive integrated moving average models and other related forecast procedures. *International Statistical Review, 54*(1), 51–66.

Durbin, J. (1986). Structural Modelling for Time Series Analysis. Shirley Kalleck Memorial Lecture, United States Bureau of the Census Second Annual Research Conference.

Gelb, A. ed. (1979). *Applied Optimal Estimation.* The Analytic Sciences Corporation, MIT Press, Cambridge, Mass.

Gersch, W., and G. Kitagawa. (1983). The prediction of time series with trends and seasonalities. *Journal of Business and Economic Statistics, 1*(3), 253–264.

Harrison, P. J. (1967). Exponential smoothing and short-term forecasting. *Management Science, 13*(11), 821–842.

Harrison, P. J., and C. F. Stevens (1976). Bayesian forecasting. *Journal of the Royal Statistical Society, B, 38*, 205–247.

Harvey, A. C., and P. H. J. Todd. (1983). Forecasting economic time series with structural and Box–Jenkins models: A case study. *Journal of Business and Economic Statistics, 1*(4), 299–307.

Hunter, J. S. (1986). The exponentially weighted moving average. *Journal of Quality Technology, 18*(4), 203–209.

Maravall, A. (1985). On structural time series models and the characterization of components. *Journal of Business and Economic Statistics, 3*(4), 350–355.

Sweet, A. (1986). Control charts using coupled exponentially weighted moving averages. *IIE Transactions, March*, 26–33.

Takata, S., M. Ogawa, P. Bertok, J. Ootsuka, K. Matushima, and T. Sata. (1985). Real-time monitoring system of tool breakage using Kalman filtering. *Robotics & Computer-Integrated Manufacturing, 2*(1), 33–40.

7
An Application of Adaptive Kalman Filtering to Statistical Process Control

Stephen V. Crowder
Corning Glass Works
Corning, New York

1. INTRODUCTION

1.1 Background and the Kalman Model

The Kalman filter originally appeared in the engineering literature in Kalman (1960) and Kalman and Bucy (1961). Since its introduction, it has been commonly used by control engineers and other physical scientists in such areas as missile trajectory and satellite orbit estimation.

The Kalman filter has also appeared extensively in the econometrics literature. Harrison (1967) uses a special case of the model as a tool in short-term sales forecasting. Sarris (1973) addresses the problem of estimating time-varying regression coefficients from a Bayesian point of view, and Sant (1977) applies generalized least squares to the same problem.

Until recently, however, the Kalman filter had not appeared in the statistical literature. Harrison and Stevens (1971, 1976) derived the Kalman filter from a Bayesian forecasting point of view. Duncan and Horn (1972) demonstrated the equivalence of Kalman filter theory and random parameter linear regression theory. While the original development of the Kalman filter is in a language foreign to statisticians, papers such as these have illustrated the filter's relation to linear models of regression and time series analysis. And because of its usefulness in applications, the Kalman filter is being viewed with increasing interest by statisticians.

Morrison and Pike (1977) applied the Kalman model to derive a short-term forecasting algorithm with time-varying parameters. Ledolter (1979) used Kalman filtering techniques to calculate recursive estimates in regression and *autoregressive integrated moving average* (ARIMA) time series models. Sallas and Harville (1981) extended the Kalman model to derive recursive estimation equations for mixed models.

The derivation of the Kalman filter given here will follow that of Harrison and Stevens (1971).

The Kalman model can be described by the matrix equations

$$Y_t = F_t\theta_t + \varepsilon_t \qquad t = 1, 2, \ldots \tag{1}$$

$$\theta_t = G_t\theta_{t-1} + v_t \tag{2}$$

where F_t and G_t are known. The data at time t are represented by Y_t, which may be either scalar or vector valued. The dependence of Y_t on θ_t, the unobservable state of nature, is described in Eq. (1), the observation equation. The observation error ε_t is assumed to be normally distributed with mean zero and known variance V_t. The state of nature θ_t is assumed to change over time according to Eq. (2), the system equation. The matrix G_t describes the transition of the state of nature from time $t - 1$ to t. The system equation error v_t is normally distributed with mean zero and known variance W_t. One also assumes that ε_t and v_t are independent. Note that the matrices F_t and G_t of Eqs. (1) and (2) may change over time, as also may the matrices V_t and W_t.

1.2 The Kalman Filter

The Kalman filter is a recursive procedure for inference concerning the state of nature θ_t in Eq. (2). The derivation given here uses a Bayesian approach to estimating θ_t given the data through time t. Represent the data available through time t as

$$Y^t = (Y_1, Y_2, \ldots, Y_t)$$

At time $t - 1$ one's knowledge of θ_{t-1} can be expressed in terms of a posterior distribution for θ_{t-1} given Y^{t-1},

$$\theta_{t-1} | Y^{t-1} \sim N(\hat{\theta}_{t-1}, Q_{t-1}) \tag{3}$$

where $\hat{\theta}_{t-1}$ and Q_{t-1} are the mean and variance of $\theta_{t-1} | Y^{t-1}$. Prior to observing Y_t, from Eq. (2) our knowledge of θ_t can be described by the conditional distribution

$$\theta_t | Y^{t-1} \sim N(G_t \hat{\theta}_{t-1}, R_t) \tag{4}$$

where $R_t = G_t Q_{t-1} G_t' + W_t$. Now, let e_t denote the error in predicting Y_t given data through time $t - 1$:

$$e_t = Y_t - \hat{Y}_t = Y_t - F_t G_t \hat{\theta}_{t-1} \tag{5}$$

Since F_t, G_t, and $\hat{\theta}_{t-1}$ are all known, observing Y_t is equivalent to observing e_t. Since $\varepsilon_t \sim N(0, V_t)$ we have from Eq. (1) that

$$\begin{aligned} e_t &= Y_t - F_t G_t \hat{\theta}_{t-1} \\ &= F_t(\theta_t - G_t \hat{\theta}_{t-1}) + \varepsilon_t \end{aligned}$$

and

$$e_t | \theta_t, Y^{t-1} \sim N[F_t(\theta_t - G_t \hat{\theta}_{t-1}), V_t] \tag{6}$$

Using a standard result from Anderson (1971) and the distributions for $\theta_t | Y^{t-1}$ and $e_t | \theta_t, Y^{t-1}$ we have that

$$\left. \begin{bmatrix} \theta_t \\ e_t \end{bmatrix} \right| Y^{t-1} \sim N \left\{ \begin{bmatrix} G_t \hat{\theta}_{t-1} \\ 0 \end{bmatrix}, \begin{bmatrix} R_t & R_t F_t' \\ F_t R_t & V_t + F_t R_t F_t' \end{bmatrix} \right\} \tag{7}$$

Now, using Eq. (7) and conditioning on e_t, the distribution of $\theta_t | Y^t$ is equivalent to the distribution of $\theta_t | e_t, Y^{t-1}$, namely,

$$\theta_t | Y^t \sim N(\hat{\theta}_t, Q_t) \tag{8}$$

where

$$\hat{\theta}_t = G_t\hat{\theta}_{t-1} + R_tF_t'(V_t + F_tR_tF_t')^{-1}e_t$$
$$Q_t = R_t - R_tF_t'(V_t + F_tR_tF_t')^{-1}F_tR_t \quad (9)$$

and

$$e_t = Y_t - F_tG_t\hat{\theta}_{t-1}$$

again, using standard results of multivariate analysis. So the cycle from Eq. (3) to Eq. (8) is completed, and the procedure moves on to time $t + 1$. The recursive procedure described by Eq. (9) is begun at time zero by choosing $\hat{\theta}_0$ and Q_0 to be best guesses about the mean and the variance of θ_0, respectively.

Note in Eq. (9) that the posterior mean of $\theta_t | Y^t$ is the sum of two quantities, $G_t\hat{\theta}_{t-1}$, and a multiple of the one-step-ahead forecast error e_t. The term $G_t\hat{\theta}_{t-1}$ is the mean of the prior distribution of $\theta_t | Y^{t-1}$, and e_t is the error in predicting Y_t given the data through time $t - 1$. Therefore, we can think of the Kalman filter as an updating procedure that combines a best prior guess about θ_t with a correction factor depending on e_t, a measure of how useful that prior guess has been in predicting the current observation.

Here we have assumed normality for the error vectors ε_t and v_t. The estimator $\hat{\theta}_t$ in Eq. (9), which is the Kalman filter estimate, is the posterior mean for θ_t given Y^t and is thus the Bayes estimator under squared error loss. Duncan and Horn (1972) show that even if the error vectors are not normally distributed, the Kalman filter estimator will still be the minimum mean square linear estimator provided the ε_t and v_t are independent vectors with means zero and respective variances V_t and W_t.

Harrison and Stevens (1976) present several examples illustrating that the Bayesian approach to forecasting includes many conventional methods such as linear regression, exponential smoothing, and linear time series models as special cases. One of the advantages of the Kalman filter approach to estimation and forecasting is the parametric structure of the model. At any given time, probabilistic information on the parameters is available in the form of a posterior distribution given all available data. Also, the sequential model definition [Eqs. (1, 2)] describes how the parameters change in time, both systematically and as a result of random shocks.

The recursive nature of the algorithm (9) is important. In estimating and forecasting practice it means that the current posterior distribution of θ_t may be calculated from the most recent observation Y_t, the posterior distribution of $\theta_{t-1}|Y^{t-1}$, and the current observation and system variances. Thus, it is not necessary to maintain the entire process history, increasing the Kalman filter's ease of use and computational efficiency.

1.3 Applications of the Kalman Filter to Process Control

The use of the Kalman filter as an estimation technique in statistical process control is increasing. In cases where prior information about the process is available, procedures based on the Kalman filter can be superior to the classical procedures like Shewhart and CUSUM control charts. Pike et al. (1978) apply the Kalman filter to special nuclear materials control and accountability and show it to be superior to the classical approaches. Phadke (1981) uses the Kalman filter as a tool in quality auditing. MacGregor (1973) and MacGregor and Wong (1980) use the Kalman filter to develop stochastic control theory for chemical process application.

Here we will use a generalization of the steady model of Harrison and Stevens (1976), a special case of the Kalman filter model [Eqs. (1, 2)], to develop an adaptive algorithm for monitoring a univariate process mean. The properties of the algorithm will be compared via simulation to the commonly used Shewhart-type monitoring of a process mean under the classical "stable process" *independent and identically distributed random variables* (iid) models.

2. AN ADAPTIVE KALMAN FILTER FOR A PROCESS MEAN

2.1 Classical Evaluation of Control Procedures by Examination of Run Length Properties Under IID Models

Duncan (1974) gives a nice discussion of Shewhart-type variables control charts, including the X-bar chart, range chart, and s^2 chart. Using the Shewhart charting approach, a statistic Q_t, say, is computed from data collected at time t. If the Q_t value is larger than a specified upper control limit U, say, or smaller than a lower control limit L, say, the process is deemed "out of control." In this case, an investigation is undertaken to find the cause of the extreme value.

For the X-bar chart, with $\bar{X}_t$ the sample mean of n observations at time t,

$$Q_t = \bar{X}_t$$
$$L = \text{target} - 3\sigma(\bar{X}_t)$$
$$U = \text{target} + 3\sigma(\bar{X}_t)$$

where $\sigma(\bar{X}_t)$ is the standard deviation of $\bar{X}_t$. This Shewhart procedure is often justified using an assumption that the X_i terms are IID $N(\mu, \sigma^2)$.

One way to quantify what any control procedure will do is to find the mean number of samples required to get an out-of-control signal assuming that the Q_t terms are IID with some known distribution. If T is the number of samples before a Q_t value first plots out of control, then the *average run length* (ARL) is the mean of T under an IID model. A vast literature describing Shewhart-type charts and their properties exists. See for example, Roberts (1966) and Duncan (1974).

2.2 A Generalization of the Steady Model

Consider the model

$$y_{it} = \theta_t + \varepsilon_{it} \qquad i = 1, 2, \ldots, n$$
$$\theta_t = \theta_{t-1} + v_t \quad t = 1, 2, \ldots \tag{10}$$

where θ_t is the true, unknown process mean at time t, $\{y_{it}, i = 1, 2, \ldots, n\}$ are observed scalar outputs at time t, each with mean θ_t, and $\{\varepsilon_{it},\ i = 1, 2, \ldots, n;\ t = 1, 2, \ldots\}$ and $\{v_t,\ t = 1, 2, \ldots\}$ are independent sequences of independent normal random variables with zero mean and variances

$$\operatorname{var}(\varepsilon_{it}) = \sigma_{\varepsilon t}^2 \qquad i = 1, 2, \ldots, n$$
$$\operatorname{var}(v_t) = \sigma_{vt}^2 \qquad t = 1, 2, \ldots \tag{11}$$

This model is a generalization of the steady model, with variance terms which can change over time. Again, this is a special case of the Kalman filter model [Eqs. (1), (2)] with

$$F_t \equiv \mathbf{1}_n, \qquad \text{an } n\text{-vector of ones}$$

$$V_t = \sigma_{\varepsilon t}^2 \mathbf{I}_n, \qquad \mathbf{I}_n \; n \times n \text{ identity matrix,}$$

$$G_t \equiv 1$$

$$W_t = \sigma_{vt}^2$$

The y_{it} terms $(i = 1, 2, \ldots, n)$ could represent repeated measurements of the same article sampled at time t, or measurements of different articles sampled during a short interval about time t.

Note that this model extends the classical Shewhart IID model by allowing the process mean to wander over time. If $\sigma_{vt}^2 = \sigma_v^2$ for every t, the parameter sequence θ_t is a simple random walk. This model is appropriate for a situation where the most important characteristic of the process in question is its current mean level, with persistent growth or decline either absent or unimportant. Harrison and Stevens (1976) discuss this model as a special case of their dynamic linear model. Harrison (1967) uses the model to describe customer demand for a steady-selling product, and as a tool in short-term forecasting. Meinhold and Singpurwalla (1983) discuss the model as an example in their overview of the Kalman filter. Also, the sample means from model [Eqs. (10), (11)] have the same autocorrelation structure as a class of ARIMA(0, 1, 1) models of Box and Jenkins (1970), models that are widely used in practice. Box and Jenkins mention that models of this kind have often been found useful in inventory control problems, in econometrics, and in representing some kinds of disturbance occurring in industrial processes. In the glass–ceramics industry, this model has been used to monitor the drift in certain physical properties of incoming batch raw materials.

2.3 The Associated Kalman Filter

The algorithm for monitoring θ_t of this model is derived as before. Letting $\mathbf{y}_t$ represent the vector of outputs observed at time t and y^t represent the set of all outputs observed through time t, let $\hat{\theta}_{t-1}$ and q_{t-1} be the mean and variance of $\theta_{t-1} | y^{t-1}$. Then

$$\theta_{t-1} | y^{t-1} \sim N(\hat{\theta}_{t-1}, q_{t-1})$$

and this distribution represents our knowledge about θ_{t-1} at time $t-1$. Using the same argument as in Section 1, the posterior distribution for θ_t after time t is

$$\theta_t | y^t \sim N(\hat{\theta}_t, q_t)$$

where

$$\hat{\theta}_t = (1 - k_t)\hat{\theta}_{t-1} + k_t \bar{y}_t$$

$$k_t = \left(\frac{q_{t-1} + \sigma_{vt}^2}{q_{t-1} + \sigma_{vt}^2 + \sigma_{\varepsilon t}^2/n}\right)$$

$$q_t = (q_{t-1} + \sigma_{vt}^2)\left(\frac{\sigma_{\varepsilon t}^2/n}{\sigma_{\varepsilon t}^2/n + \sigma_{vt}^2 + q_{t-1}}\right) \tag{12}$$

These are the Kalman filter recursive equations for estimating the θ_t sequence, under the model described by Eqs. (10) and (11). If all the variance terms are known, then $\hat{\theta}_t$ is a weighted moving average, depending on the data only through $\bar{y}_1, \ldots, \bar{y}_t$. The Kalman weights k_t can be calculated independent of the data, using Eq. (12). If in addition the variance terms are constant, the sequence k_t converges to a steady-state limit (hence the name steady model), and $\hat{\theta}_t$ is an exponentially weighted moving average.

2.4 Estimation of Variance Components

We will now address the problem of unknown variance components that must be estimated from the data. In this case, the incorporation of the estimated variance components into the algorithm for estimating the process mean produces what is known as an adaptive Kalman filter.

For the case of constant but unknown variance components, several adaptive filtering algorithms have been proposed in the engineering literature. A review and a bibliography are given by Mehra (1972). Sarris (1973) presents a maximum likelihood approach, and Louv (1984) proposes a MINQUE algorithm for estimating variance components. The approach taken here will be to use maximum likelihood to estimate variance terms *changing* over time. For the resulting adaptive filter, a proposed control charting technique will be studied.

An examination of the full likelihood of y^t at time t indicates that its form is too complicated for practical use in estimating the current variance components $\sigma^2_{\varepsilon t}$ and σ^2_{vt}. A second approach is to use the likelihood of successive differences $\mathbf{y}_t - \mathbf{y}_{t-1}$ in the estimation. However, the resulting estimators are unacceptable in that they depend on the particular ordering of the data $\{y_{it}, i = 1, \ldots, n\}$ in the vector $\mathbf{y}_t$.

To overcome the lack of symmetry that results from this approach and to obtain estimators of $\sigma^2_{\varepsilon t}$ and σ^2_{vt} that are symmetric in the data, we will instead consider only the joint likelihood function of the following three independent statistics.

Let

$$z_t = \bar{y}_t - \bar{y}_{t-1}$$

be the difference in successive sample means,

$$d_t = \sum_{i=1}^{n} (y_{it} - \bar{y}_t)^2$$

and

$$d_{t-1} = \sum_{i=1}^{n} (y_{it-1} - \bar{y}_{t-1})^2$$

Then the joint likelihood function of (z_t, d_{t-1}, d_t) for $\sigma^2_{vt} \geqslant 0$, $\sigma^2_{\varepsilon t-1} > 0$, and $\sigma^2_{\varepsilon t} > 0$ can be expressed

$$L(z_t, d_{t-1}, d_t; \sigma^2_{vt}, \sigma^2_{\varepsilon t-1}, \sigma^2_{\varepsilon t})$$

$$= \left[2\pi\left(\sigma^2_{vt} + \frac{\sigma^2_{\varepsilon t-1}}{n} + \frac{\sigma^2_{\varepsilon t}}{n}\right)\right]^{-1/2} \exp\left[\frac{-z_t^2}{2(\sigma^2_{vt} + \sigma^2_{\varepsilon t-1}/n + \sigma^2_{\varepsilon t}/n)}\right]$$
$$\times c_n \frac{d_{t-1}^{(n-1)/2-1}}{(\sigma^2_{\varepsilon t-1})^{(n-1)/2}} \exp\left(\frac{-d_{t-1}}{2\sigma^2_{\varepsilon t-1}}\right) c_n \frac{d_t^{(n-1)/2-1}}{(\sigma^2_{\varepsilon t})^{(n-1)/2}} \exp\left(\frac{-d_t}{2\sigma^2_{\varepsilon t}}\right) \tag{13}$$

Note that the statistics used to form Eq. (13) do not depend on the ordering of the data collected at time $t - 1$ or time t.

The likelihood function above can be maximized over nonnegative choices of σ^2_{vt} and positive choices of $\sigma^2_{\varepsilon t-1}$ and $\sigma^2_{\varepsilon t}$ to obtain estimates of σ^2_{vt} and $\sigma^2_{\varepsilon t}$, the variances components of interest at time

t, for use in the recursions (12). Notice that

$$\log L = c_0 - \tfrac{1}{2}\log\left(\sigma_{vt}^2 + \frac{\sigma_{\varepsilon t-1}^2}{n} + \frac{\sigma_{\varepsilon t}^2}{n}\right)$$

$$- \frac{z_t^2}{2(\sigma_{vt}^2 + \sigma_{\varepsilon t-1}^2/n + \sigma_{\varepsilon t}^2/n)}$$

$$- \left(\frac{n-1}{2}\right)\log(\sigma_{\varepsilon t-1}^2) - \frac{d_{t-1}}{2\sigma_{\varepsilon t-1}^2}$$

$$- \left(\frac{n-1}{2}\right)\log(\sigma_{\varepsilon t}^2) - \frac{d_t}{2\sigma_{\varepsilon t}^2}$$

Differentiating with respect to σ_{vt}^2,

$$\frac{\partial(\log L)}{\partial\sigma_{vt}^2} = \frac{1}{2(\sigma_{vt}^2 + \sigma_{\varepsilon t-1}^2/n + \sigma_{\varepsilon t}^2/n)} \times \left[\frac{z_t^2}{(\sigma_{vt}^2 + \sigma_{\varepsilon t-1}^2/n + \sigma_{\varepsilon t}^2/n)} - 1\right] \qquad (14)$$

So, $\partial(\log L)/\partial\sigma_{vt}^2 = 0$ when $\sigma_{vt}^2 = z_t^2 - (\sigma_{\varepsilon t-1}^2 + \sigma_{\varepsilon t}^2)/n$.

From the form of Eq. (14), for any fixed positive $\sigma_{\varepsilon t-1}^2$ and $\sigma_{\varepsilon t}^2$, L is maximized as a function of $\sigma_{vt}^2 \geqslant 0$ by

$$\hat{\sigma}_{vt}^2 = \max[0, z_t^2 - (\sigma_{\varepsilon t-1}^2 + \sigma_{\varepsilon t}^2)/n]$$

To find overall maximizers of Eq. (13) it then suffices to separately maximize (numerically) $\log L(z_t, d_{t-1}, d_t;\ 0,\ \sigma_{\varepsilon t-1}^2, \sigma_{\varepsilon t}^2)$ and $\log L(z_t, d_{t-1}, d_t;\ z_t^2 - (\sigma_{\varepsilon t-1}^2 + \sigma_{\varepsilon t}^2)/n,\ \sigma_{\varepsilon t-1}^2, \sigma_{\varepsilon t}^2)$ over positive choices of $(\sigma_{\varepsilon t-1}^2,\ \sigma_{\varepsilon t}^2)$ and compare the resulting maxima. Then given $(\hat{\sigma}_{\varepsilon t-1}^2, \hat{\sigma}_{\varepsilon t}^2)$,

$$\hat{\sigma}_{vt}^2 = \max\{0, (\bar{y}_t - \bar{y}_{t-1})^2 - (\hat{\sigma}_{\varepsilon t-1}^2 + \hat{\sigma}_{\varepsilon t}^2)/n\} \qquad (15)$$

In practice, $\hat{\sigma}_{\varepsilon t}^2$ turns out to be approximately equal to s_t^2, the sample variance of the subgroup observed at time t.

2.5 The Adaptive Algorithm

Using the maximum likelihood estimates described in Section 2.4, the adaptive Kalman filter recursive equations become

$$\hat{\hat{\theta}}_t = (1 - \hat{k}_t)\hat{\hat{\theta}}_{t-1} + \hat{k}_t\bar{y}_t$$

$$\hat{k}_t = \left(\frac{\hat{q}_{t-1} + \hat{\sigma}^2_{vt}}{\hat{q}_{t-1} + \hat{\sigma}^2_{vt} + \hat{\sigma}^2_{\varepsilon t}/n} \right)$$

$$\hat{q}_t = (\hat{q}_{t-1} + \hat{\sigma}^2_{vt}) \left(\frac{\hat{\sigma}^2_{\varepsilon t}/n}{\hat{\sigma}^2_{\varepsilon t}/n + \hat{\sigma}^2_{vt} + \hat{q}_{t-1}} \right) \tag{16}$$

Since the variance terms are estimated from the data, the posterior variance $\hat{q}_t$ and the Kalman weight $\hat{k}_t$ cannot be expected to converge to a limiting value. The term $\hat{\hat{\theta}}_t$ is thus an adaptive exponentially weighted moving average.

Adaptive exponentially smoothed forecasting techniques have been extensively studied, with the primary concern being how to choose and continually adjust the weighting factor k_t. One approach uses the behavior of past data in order to arrive at an optimum value, as in Rao and Shapiro (1970). Another approach, discussed by Trigg (1964), is to vary the weighting factor according to the value of a tracking signal that depends on previous forecasting errors.

In Eq. (12) the weighting factor k_t is determined from the Kalman filter recursions and knowledge of the variance terms. In Eq. (16), the variance terms that determine k_t are unknown, so maximum likelihood estimates are used to obtain an estimate of k_t, $\hat{k}_t$. In either case, choice of weighting factor is a natural consequence of the Kalman structure.

Consider the weight $\hat{k}_t$ in Eq. (16). During periods when the process mean is changing the most, a large weighting factor is desirable, to increase the influence of the most recent samples on $\hat{\hat{\theta}}_t$. If the estimate $\hat{\sigma}^2_{vt}$ in Eq. (16) is large relative to $\hat{q}_{t-1}$ and $\hat{\sigma}^2_{\varepsilon t}$, then $\hat{k}_t$ is large, and the most recent sample mean has the most influence. From the form of the estimator $\hat{\sigma}^2_{vt}$ in Eq. (15), we see that a large value of $(\bar{y}_t - \bar{y}_{t-1})^2$ will result in a large estimate of σ^2_{vt} and hence a large weighting factor. So, if the process mean is changing significantly, it will result in large values for $(\bar{y}_t - \bar{y}_{t-1})^2$ and $\hat{\sigma}^2_{vt}$, as desired. During periods when the process is relatively stable, a small value for the weighting factor is desirable. In this case, small values of $(\bar{y}_t - \bar{y}_{t-1})^2$ will typically be observed, and the estimate of $\hat{\sigma}^2_{vt}$ should be close to or equal zero. Then, the estimate $\hat{k}_t$ will also be small, and the current data will have less influence on $\hat{\hat{\theta}}_t$, as desired.

Here we are concerned with the application of this adaptive estimation technique in control charting to monitor a process mean. We will consider a type of adaptive signaling criterion, proposed by Hoadley (1981), that takes advantages of the posterior distribution and estimated posterior variances $\hat{q}_t$.

3. CONTROL CHARTING FOR A PROCESS MEAN

3.1 Box and Whisker Plot

From Eq. (16), after the data at time t have been observed, we have estimates of the posterior mean and variance of θ_t, the current process mean. The posterior mean and variance of θ_t given y^t are estimated, respectively, by

$$\hat{\hat{\theta}}_t = \hat{E}(\theta_t|y^t) = (1 - \hat{k}_t)\hat{\hat{\theta}}_{t-1} + \hat{k}_t\bar{y}_t$$

and

$$\hat{q}_t = \hat{V}(\theta_t|y^t) = (\hat{q}_{t-1} + \hat{\sigma}_{vt}^2)\left(\frac{\hat{\sigma}_\varepsilon^2/n}{\hat{\sigma}_\varepsilon^2/n + \hat{\sigma}_{vt}^2 + q_{t-1}}\right)$$

Using an idea suggested by Hoadley (1981), a box and whisker plot can be made each period, in addition to calculation of the adaptive point estimate $\hat{\hat{\theta}}_t$. The box plot will be used to make a graphical representation of the estimated posterior distribution of the current process mean θ_t. An "O" at the middle of the box will represent the estimated posterior mean of the process. A "Y" on the chart at time t will represent the observed sample mean at time t. The top and bottom of the box and the end of the whiskers will be drawn a fixed number of estimated posterior standard deviations of θ_t from $\hat{\hat{\theta}}_t$. When either the top or bottom whisker fails to cross the target value, the process is deemed "out of control." A warning signal is given if the box fails to cross the target line. An illustration of the charting procedure is given in Figure 1.

In Figure 1, the box and whiskers could extend, respectively, 2 and 3 posterior standard deviations from $\hat{\hat{\theta}}_t$. The box plot at left represents a process in control, the center box plot signals a warning that the process mean may be wandering off target, while

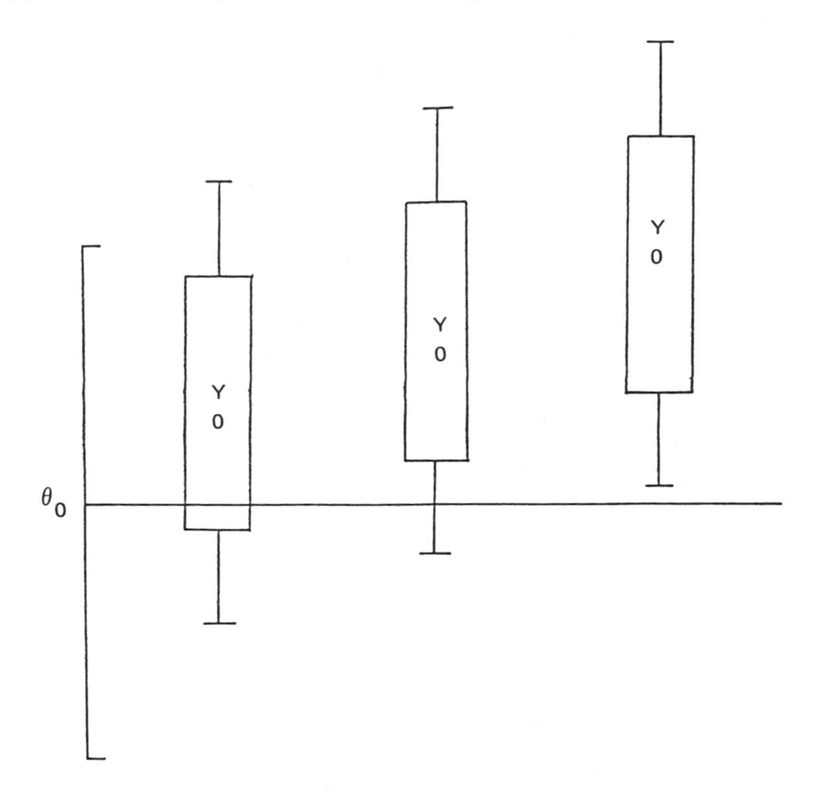

FIGURE 1 Box and whisker plots for a process mean: θ_0 is target value, O is estimate of process mean, θ_t, and Y is sample mean, $\bar{y}_t$.

the right-hand box plot signals that the process is out of control. Figure 2 illustrates the adaptive algorithm at work in box plot form, using simulated $(\bar{y}_t, s_t^2)$ pairs, for $t = 1, 2, \ldots, 10$.

3.2 Properties of the Algorithm and a Comparison to the Shewhart Approach

If the model of Eqs. (10) and (11) adequately describes the process of interest, then $\hat{\hat{\theta}}_t$ of Eq. (16) should effectively track the true process level θ_t. The question remains as to the performance of the procedure if in fact the standard Shewhart IID model adequately describes the process.

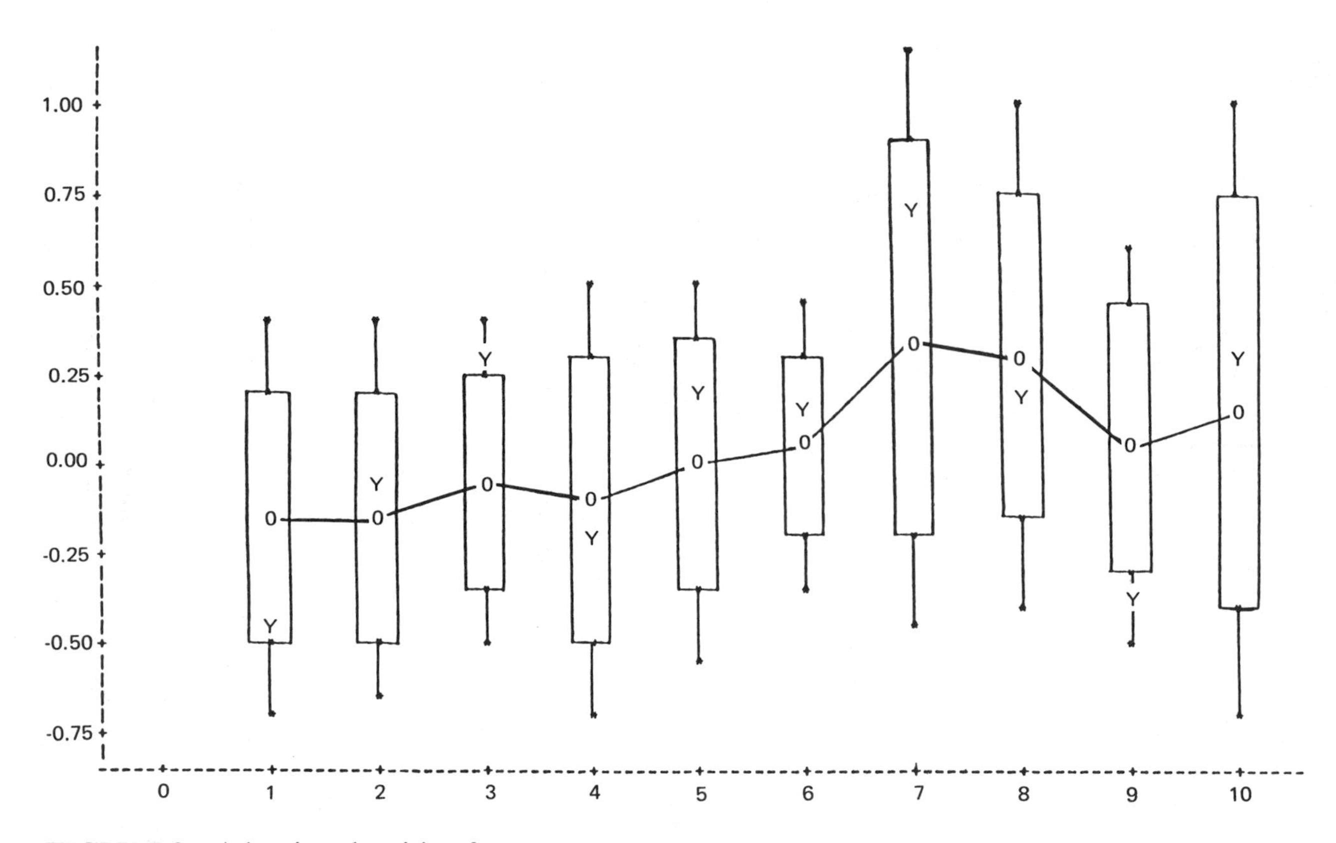

FIGURE 2 Adaptive algorithm for process mean: O is estimate of posterior mean, Y is sample mean. Box and whiskers extend 2 and 3 estimated posterior standard deviations, respectively, from O.

ARL properties of the control scheme based on the adaptive Kalman filter were studied, using an IID model for the variables $\{y_{it}, i = 1, \ldots, n; t = 1, 2, \ldots\}$. Because of the adaptive nature of the procedure, analytical expressions for the ARLs could not be derived. Simulations were performed to obtain empirical estimates of the ARLs. Table 1 contains these ARLs for the adaptive control scheme, for subgroup samples of size $n = 5$. The ARLs are based on 1000 simulated runs of an IID process. The standard errors of the simulation estimates are less than 5% of the estimated means.

In Table 1, L refers to the box plot control limit in terms of multiples of $\sqrt{\hat{q}_t}$. The column heading D refers to the true fixed deviation from nominal, measured in units of the standard deviation σ. The algorithms were started with $\hat{\hat{\theta}}_0 = 0$ and $\hat{q}_0 = \sigma^2$, the variance of the y_{it} terms from the IID model. Our box and whisker approach corresponds to the plotting of "Bayesian" t statistics since we are essentially plotting values of $\hat{\hat{\theta}}_t/\sqrt{\hat{q}_t}$.

The approach described above can be compared to a Shewhart

TABLE 1 ARL Values for Adaptive Filter with Box Plot Limits, $n = 5$

	L, number of posterior standard deviations							
D^a	3	4	5	6	7	8	9	10
0.00	24.3	62.8	140.0	246.5	471.5	782.3	1177.4	1795.6
0.25	13.6	31.0	71.3	139.4	250.4	418.7	674.9	1014.7
0.50	5.9	10.6	21.2	39.9	77.2	135.9	231.9	388.4
0.75	3.3	5.6	8.7	15.5	24.7	42.0	70.5	116.5
1.00	2.3	3.3	4.8	7.4	11.2	17.2	26.2	39.6
1.50	1.4	1.9	2.5	3.4	4.5	6.3	8.2	10.8
2.00	1.2	1.4	1.7	2.2	2.7	3.3	4.2	5.2
2.50	1.0	1.2	1.4	1.6	1.9	2.3	2.8	3.3
3.00	1.0	1.1	1.2	1.4	1.5	1.8	2.1	2.4

[a]Deviation from nominal, measured in units of σ.

t statistic control chart, in which successive values of

$$Q_t = \sqrt{n}\bar{y}_t/s_t$$

are plotted, where $\bar{y}_t$ and s_t are the sample mean and sample standard deviation, respectively, observed at time t. Under an IID normal model, $\bar{y}_t \sim N(\mu, \sigma^2/n)$, $(n-1)s_t^2/\sigma^2 \sim \chi^2_{(n-1)}$, and Q_t has a noncentral t distribution with noncentrality parameter $\sqrt{n}\mu/\sigma$. Table 2 gives the exact ARLs of the t statistic procedure, used for comparison. The two procedures can be compared by fixing in-control ARLs at a particular value and comparing ARLs over the out-of-control region of interest. For example, with in-control ARLs set at 500, Table 3 gives the out-of-control ARL values that result, obtained using interpolation in Tables 1 and 2.

From Table 3 we see that the adaptive control procedure appears to be superior to the standard t statistic procedure in terms of ARLs corresponding to small shifts in the process mean. For large shifts in the process mean, the ARLs corresponding to Shewhart plotting of the t statistic are similar to those corresponding to box and whisker plotting of the adaptive filter estimate.

TABLE 2 ARL Values for t Statistic Control Procedure, $n = 5$

	L, t statistic control limit							
D^a	3	4	5	6	7	8	9	10
0.00	25.0	62.0	133.5	257.6	456.2	755.3	1185.1	1779.1
0.25	17.1	40.4	85.0	161.5	283.3	466.1	728.0	1089.4
0.50	8.6	18.9	38.2	70.9	122.5	199.4	309.1	460.0
0.75	4.6	9.5	18.3	33.1	56.1	90.3	138.8	205.3
1.00	2.9	5.4	9.9	17.2	28.7	45.5	69.2	101.7
1.50	1.5	2.4	3.9	6.3	9.9	15.2	22.6	32.6
2.00	1.1	1.5	2.1	3.1	4.6	6.8	9.8	13.7
2.50	1.0	1.2	1.5	2.0	2.7	3.7	5.2	7.1
3.00	1.0	1.0	1.2	1.4	1.8	2.4	3.2	4.2

[a]Deviation from nominal, measured in units of σ.

TABLE 3 ARL Comparison of t Statistic Plotting and Adaptive Filter Box Plot Procedure with In-Control ARL = 500

	Control Limits	
D^a	t statistic $L = 7.2$	box plot $L = 7.1$
0.00	500.0	500.0
0.25	310.0	266.0
0.50	134.0	82.0
0.75	61.0	26.3
1.00	31.0	11.8
1.50	10.7	4.7
2.00	4.9	2.8
2.50	2.9	1.9
3.00	1.9	1.5

[a]Deviation from nominal, measured in units of σ.

3.3 Comments on Control Chart Design and Implementation

The performance of the adaptive filter has been compared to the performance of t statistic plotting under an IID normal model, and shown to be superior for detecting small shifts in the process mean.

For purposes of control chart design, the user determines what magnitude of shift should be detected quickly. Table 1 can then be used to compare in-control and out-of-control ARLs for different choices of L. The choice of L that gives the most favorable ARL profile over the region of interest is then used in the box and whisker plotting procedure. That is, successive box and whisker plots with endpoints at $\hat{\hat{\theta}}_t \pm L\sqrt{\hat{q}_t}$ are then compared to a nominal or target value.

The procedure can easily be implemented for on-line monitoring given the use of a microcomputer for numerical evaluation of the adaptive weights k_t. If no such numerical support is available, the estimates

$$\hat{\sigma}_{\varepsilon t}^2 = s_t^2$$

and

$$\hat{\sigma}_{vt}^2 = \max\{0, (\bar{y}_t - \bar{y}_{t-1})^2 - (s_t^2 + s_{t-1}^2)/n\}$$

can be used to approximate the weights $\hat{k}_t$.

Smaller simulation studies by Crowder (1986) suggested that the performance of the adaptive filter for larger subgroup sample sizes is consistent with the results presented here.

REFERENCES

Anderson, T. W. (1971). *An Introduction to Multivariate Statistical Analysis.* Wiley, New York.

Box, G. E. P., and G. M. Jenkins. (1970). *Time Series Analysis, Forecasting and Control.* Holden-Day, San Francisco.

Crowder, S. V. (1986). Kalman filtering and statistical process control. Unpublished Ph.D. dissertation. Iowa State University Library, Ames, Iowa.

Duncan, A. J. (1974). *Quality Control and Industrial Statistics.* Irwin, Homewood, Illinois.

Duncan, D. B., and S. D. Horn. (1972). Linear dynamic recursive estimation from the viewpoint of regression analysis. *Journal of the American Statistical Association, 67*, 815–821.

Harrison, P. J. (1967). Exponential smoothing and short-term sales forecasting. *Management Science, 13*, 821–842.

Harrison, P. J., and C. F. Stevens. (1971). A Bayesian approach to short-term forecasting. *Operational Research Quarterly, 22*, 341–362.

Harrison, P. J. and C. F. Stevens. (1976). Bayesian forecasting. *Journal of the Royal Statistical Society, B, 38*, 205–228.

Hoadley, B. (1981). The quality measurement plan. *Bell System Technical Journal, 60*, 215–273.

Kalman, R. E. (1960). A new approach to linear filtering and prediction theory. *Journal of Basic Engineering, 82*, 34–45.

Kalman, R. E., and R. S. Bucy. (1961). New results in linear filtering and prediction theory. *Journal of Basic Engineering*, *83*, 95–108.

Ledolter, J. (1979). A recursive approach to parameter estimation in regression and time series models. *Communications in Statistics*, *A*, 1227–1246.

Louv, W. C. (1984). Adaptive filtering. *Technometrics*, *26*, 399–409.

MacGregor, J. F. (1973). Optimal discrete stochastic control theory for process application. *Canadian Journal of Chemical Engineering*, *51*, 468–478.

MacGregor, J. F., and A. K. L. Wong. (1980). Multivariate model identification and stochastic control of a chemical reactor. *Technometrics*, *22*, 453–464.

Mehra, R. K. (1972). Approaches to adaptive filtering. *IEEE Transactions on Automatic Control*, *17*, 693–698.

Meinhold, R. J., and N. D. Singpurwalla. (1983). Understanding the Kalman filter. *American Statistician*, *37*, 123–127.

Morrison, G. W., and D. H. Pike. (1977). Kalman filtering applied to statistical forecasting. *Management Science*, *23*, 768–774.

Phadke, M. S. (1981). Quality evaluation plan using adaptive Kalman filtering. *Bell System Technical Journal*, *61*, 2081–2107.

Pike, D. H., G. W. Morrison, and D. J. Downing. (1978). Time Series Analysis Applicable to Nuclear Accountability Data. ORNL/NUREG/CSD-10, Oak Ridge National Laboratory.

Rao, A. G., and A. Shapiro. (1970). Adaptive smoothing using evolutionary spectra. *Management Science*, *17*, 208–218.

Roberts, S. W. (1966). A comparison of some control chart procedures. *Technometrics*, *8*, 411–430.

Sallas, W. M., and D. A. Harville. (1981). Best linear recursive estimation for mixed linear models. *Journal of the American Statistical Association*, *76*, 860–869.

Sant, D. T. (1977). Generalized least squares applied to time-varying parameter models. *Annals of Economic and Social Measurement*, *6*, 301–311.

Sarris, A. H. (1973). A Bayesian approach to estimation of time-varying regression coefficients. *Annals of Economic and Social Measurement*, *2*, 501–523.

Trigg, D. W. (1964). Monitoring a forecasting system. *Operational Research Quarterly*, *15*, 271–274.

8
A Multivariate and Stochastic Framework for Statistical Process Control

Norma Faris Hubele
Arizona State University
Tempe, Arizona

Multivariate and stochastic techniques have not been widely applied in the discrete manufacturing setting. To a large extent, this has been due to the widespread success of the Shewhart control charts and the relatively low level of sophistication of the processes being monitored. In the modern factory, however, the processes are much more complex (e.g., semiconductor manufacturing) and the cost of data collection is considerably reduced. With automation, machines can be instrumented to collect data, computers can be programmed to analyze the data, and a message can then be sent to the machine to perform the corrective action. In this environment, the potential exists for much wider application of methods that address the complexity of the processes, i.e., multivariate and stochastic techniques.

Based on a presentation madc at "Statistical Process Control: Keeping Pace with Automated Manufacturing, a National Symposium," sponsored by the Center for Professional Development and the Reliability, Availability and Serviceability Laboratory, College of Engineering and Applied Sciences, Arizona State University, November 6–7, 1986.

1. A MULTIVARIATE CONTROL FRAMEWORK

Multivariate *statistical process control* (SPC) techniques involve monitoring the quality characteristics that are interrelated. Most of the past work in multivariate SPC has focused on controlling the behavior of the mean vector and the variance–covariance matrix. Hotelling (1947) first introduced a multivariate method in his analysis of bombsight data; Alt (1985) provides a good review of the developments in this area since that time. Recent efforts have begun to address methods for monitoring individual multivariate observations (Walker, 1985).

Critical issues that must be examined by a univariate SPC schema closely parallel those requiring attention in the multivariate case. These issues are (1) detection of an out-of-control state, (2) identification of the variable, among the interrelated variables, that caused the problem, and (3) determination of the magnitude and direction of the adjustment(s) required. Specifying the acceptable (from a quality point of view) region of interrelated variable is also an issue. These points are covered by structuring our discussion around extending the notions of univariate *narrow-limit gaging* (NLG) to the multivariate case.

First developed in the United Kingdom in the 1940s (Dudding and Jennett, 1945; Ministry of Supply Advisory Service and Quality Control, 1944), NLG is a methodology designed to efficiently alert an operator to an out-of-control state or an impending one. It is typically applied to bilateral specifications: i.e., in cases where a lower and upper specification limit (LSL and USL) exist. An estimate of the process variability, σ, should be available. Normality of the process variable is required, and the difference, USL − LSL, should be greater than 6σ. The term "gaging" arises from the fact that, in the process of obtaining sample information, actual measurements need not be obtained. Only the location of the process characteristic must be noted with regard to the specification limits and the narrow gage limits. This information may be obtained through the use of gages such as those of the "go/no-go" type. Narrow or compressed gage limits are constructed a fixed number of standard deviations within the specification limits, as indicated in Figure 1. For simplicity the regions of Figure 1 are usually color-

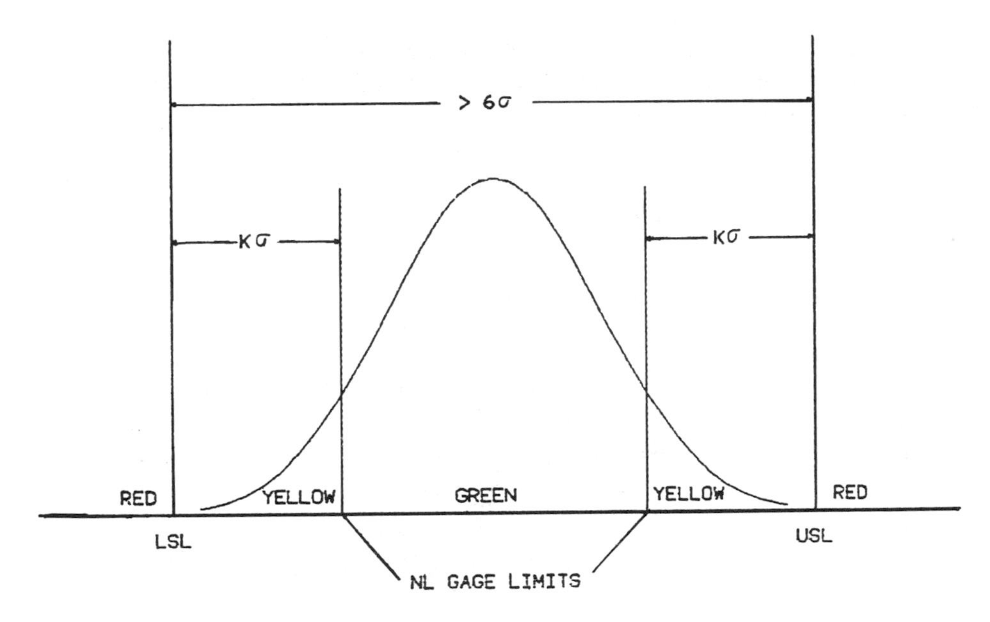

FIGURE 1 Univariate narrow-limit gaging.

coded. The compressed gage limits are used to guide the process, preventing problems rather than indicating difficulties after a problem has developed. In other words, NLG is designed to signal the approach of defective items rather than to indicate the actual presence of defectives in production. The position of the narrow gage limits is a function of risk, with the number of standard deviations (k) chosen so that, given no change in the process mean, the probability of a single yellow is rather small, but not extremely small—e.g., less than 0.20. Rules for stopping or adjusting the process have been developed both from an economic (Yu, 1983) and a risk-based (Ott, 1975) standpoint. These univariate rules will be illustrated later in the discussion on simulation in Section 1.3.

In the following discussion, we adopt a method called *multivariate narrow-limit gaging* (MNLG). We assume that the quality characteristics are interrelated and the individual observations are available for analysis in a computer-integrated environment. That is, the data and computing power are available. The appeal of univariate NLG lies in its simplicity, ease of use, and speed. In a multivariate computer-integrated environment, these benefits may not be compelling because, for example, a computer will be performing the diagnosis and adjustment. Nevertheless, the extension of NLG to the multivariate framework is intuitively appealing and raises the issues that must be addressed by even more sophisticated multivariate approaches.

1.1 Problem Definition

Let $\mathbf{x}$ be a p-component random vector with known distribution function, $F(\mathbf{x})$, defined for the parameter set $\{\lambda\}$. The random variables, $x_1, x_2, \ldots, x_p$ are assumed to be continuous with finite mean $\boldsymbol{\mu}$ and covariance matrix $\sum$. The random variables are measures of some characteristic of product quality that are important in the specification and manufacturing process. They are also assumed to be correlated; the following discussion is applicable to the case of independent random variables, however, it reduces to the univariate approach.

Assume that the design specification, S_i, is a function that defines a domain for each x_i, and call this domain DS_i. Then the probability of manufacturing a product according to design specifications is

$$P_S = \int_{\mathrm{DS}_p} \int_{\mathrm{DS}_{p-1}} \cdots \int_{\mathrm{DS}_1} dF(\mathbf{x}) \leqslant 1 \tag{1}$$

Further, assume that the narrow-limit gage function N_i, also a function of x_i, defines a domain DN_i for each x_i. Necessarily, $\mathrm{DN}_i \subseteq \mathrm{DS}_i$ for each i. Then, the probability of manufacturing a product with characteristics of product quality within the narrow gage limits is

$$P_N = \int_{\mathrm{DN}_p} \int_{\mathrm{DN}_{p-1}} \cdots \int_{\mathrm{DN}_1} dF(\mathbf{x}) \tag{2}$$

By definition, $P_N < P_S$. The problem addressed in this research is to determine DS_i and DN_i according to some acceptable limits of risk.

If $f(\mathbf{x})$, the density associated with $F(\mathbf{x})$, is multivariate normal, then

$$f(\mathbf{x}) = \frac{1}{(2\pi)^{p/2}\,|\Sigma|^{1/2}} \exp\{\tfrac{1}{2}[(\mathbf{x}-\boldsymbol{\mu})^T \Sigma^{-1} (\mathbf{x}-\boldsymbol{\mu})]\} \tag{3}$$

where Σ is the variance–covariance matrix and p is the number of variables, then it is well known that contours of constant density are ellipsoids centered at $\boldsymbol{\mu}$. Further, solid ellipsoids of $\mathbf{x}$ values satisfying

$$(\mathbf{x} - \boldsymbol{\mu})^T\Sigma^{-1}(\mathbf{x} - \boldsymbol{\mu}) \leqslant \chi_p^2(\alpha) \tag{4}$$

have probability $1 - \alpha$. Consequently, design and narrow-limit gage specifications may be defined as probability levels $P_S = 1 - \alpha_S$ and $P_N = 1 - \alpha_N$ and the domains expressed as function of equations such as Eq. (4). That is,

Region A: $\{\mathbf{x} | (\mathbf{x} - \boldsymbol{\mu})^T\Sigma^{-1}(\mathbf{x} - \boldsymbol{\mu}) \leqslant \chi_p^2(\alpha_N)\}$

Region B: $\{\mathbf{x} | (\mathbf{x} - \boldsymbol{\mu})^T\Sigma^{-1}(\mathbf{x} - \boldsymbol{\mu})$
$> \chi_p^2(\alpha_N) \cap (\mathbf{x} - \boldsymbol{\mu})^T\Sigma^{-1}(\mathbf{x} - \boldsymbol{\mu}) \leqslant \chi_p^2(\alpha_s)\}$

Region C: $\{\mathbf{x} | (\mathbf{x} - \boldsymbol{\mu})^T\Sigma^{-1}(\mathbf{x} - \boldsymbol{\mu}) > \chi_p^2(\alpha_s)\}$

For example, Figure 2 depicts a bivariate normal with

$$\mu = \begin{bmatrix} 100 \\ 200 \end{bmatrix} \qquad \Sigma = \begin{bmatrix} 100 & 112.5 \\ 112.5 & 225 \end{bmatrix}$$

$P_S = 0.999 \qquad P_N = 0.860 \qquad \rho_{x_1x_2} = 0.75$

When the parameter values μ and Σ are not known in exact terms, estimates may be used to determine the design and specification regions. Jackson (1956) pointed out that when the parameter estimates were used, then the χ_α^2 value should be replaced by Hotelling's T_α^2, where $T_\alpha^2 = [p(n-1)/(n-p)]F_{\alpha,p,n-p}$. $F_{\alpha,p,n-p}$ is obtained from the tables with significance level α, p is the dimension of the random vector $\mathbf{x}$, and n is the sample size with which the parameters are estimated. As a consequence of the law of large numbers that says $\bar{\mathbf{x}}$ converges in probability to $\boldsymbol{\mu}$ and that each sample covariance s_{ik} converges in probability to σ_{ik}, $i, k = 1, 2, \ldots, p$, so S converges in probability to Σ, we may still use the χ_α^2 distribution in some cases. Specifically, when the population is multivariate normal and both n and $n - p$ are greater than 25 or 30, then each term

$$(\mathbf{x}_j - \bar{\mathbf{x}})^TS^{-1}(\mathbf{x}_j - \bar{\mathbf{x}}) \qquad j = 1, \ldots, n$$

will behave approximately as a chi-squared random variable with p

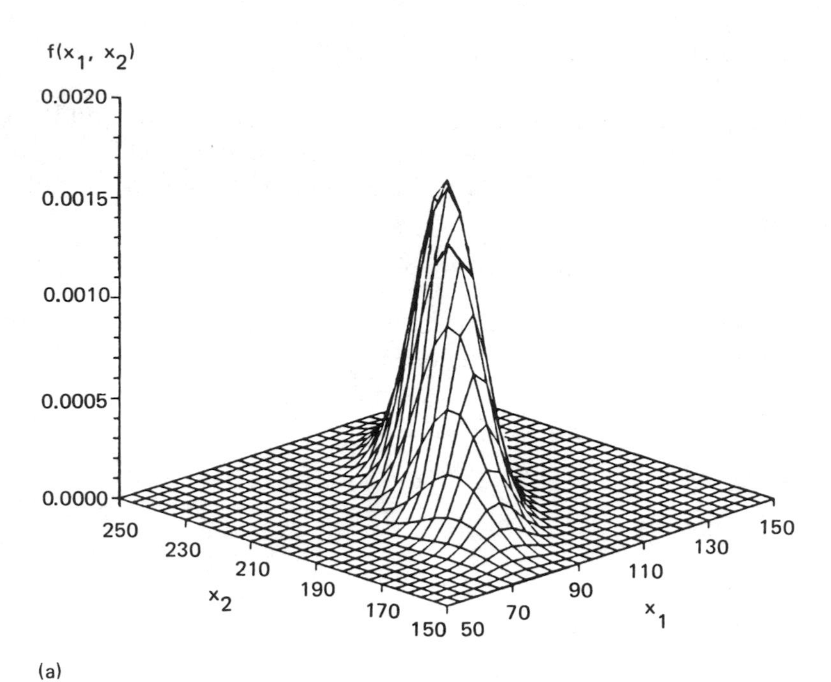

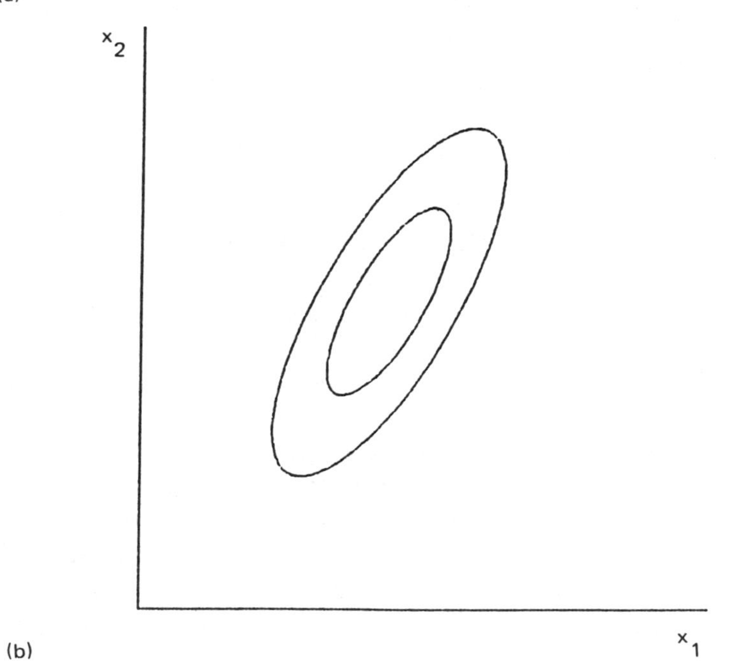

FIGURE 2 Bivariate normal distribution with (a) $\mu = \begin{bmatrix} 100 \\ 200 \end{bmatrix}$ and $\Sigma = \begin{bmatrix} 100 & 112.5 \\ 112.5 & 225 \end{bmatrix}$ and (b) 0.999 and 0.86 probability contours.

degrees of freedom. Furthermore, as one might suspect, as

$$n \to \infty, \qquad \frac{p(n-1)}{(n-p)} F_{\alpha,p,n-p} \to \chi^2_{\alpha,p}$$

In the computational results illustrated in the simulation section, the χ^2 values are used based on these above assumptions.

1.2 Rules for Adjusting the Process

Adopting the reasoning of univariate NLG, in particular, pre-control, let Region A be the acceptable ("green") region; Region B, the warning ("yellow"); and Region C, the out-of-control ("red") region (see Figure 3). These regions will indicate *when* an observation is suspect or bad. Nine subregions, determined by the intersection of straight lines formed by $\mu_1 \pm 2\sigma_{11}$, $\mu_1 \pm 4.5\sigma_{11}$, $\mu_2 \pm 2\sigma_{22}$, and $\mu_2 \pm 4.5\mu_{22}$, are superimposed on the contours to indicate *where* the problem observations occurred. Probability values for each subregion for the designated distribution are given in Figure 3. For example, the probability of a process observation occurring in subregion 1 between the elliptical contours, when there is no process change, is 0.008, or in subregion 5 (the inner ellipse) is 0.86. These regions will provide the basis for the $n = 2$ decision rules; rules for $n = 3, 4, \ldots, 10$ are direct extensions of the $n = 2$ case.

Table 1 identifies which mean(s) and/or variance(s) are to be adjusted as a result of a sequence of observations. Order of observation is not considered, and only the regions between the inner and outer ellipse are considered. Example 1: If the first observation is in subregion 8 and the second is in subregion 3, adjust μ_2 and σ_{11}. Example 2: If the first observation is in subregion 5 and the second is in subregion 6, μ_2 must be adjusted. With reference to probabilities given in Figure 3, Type I errors due to adjustment range from 0.00006 (0.008^2) to 0.00436 (0.066^2), so that a Type I error is not very likely.

The rules for $n > 2$ are a direct extension of the case where $n = 2$, and n consecutive observations are sampled. The sample observations are examined in overlapping pairs; i.e., if $n = 5$, sample pairs 1 and 2, 2 and 3, 3 and 4, and 4 and 5 are scrutinized using the

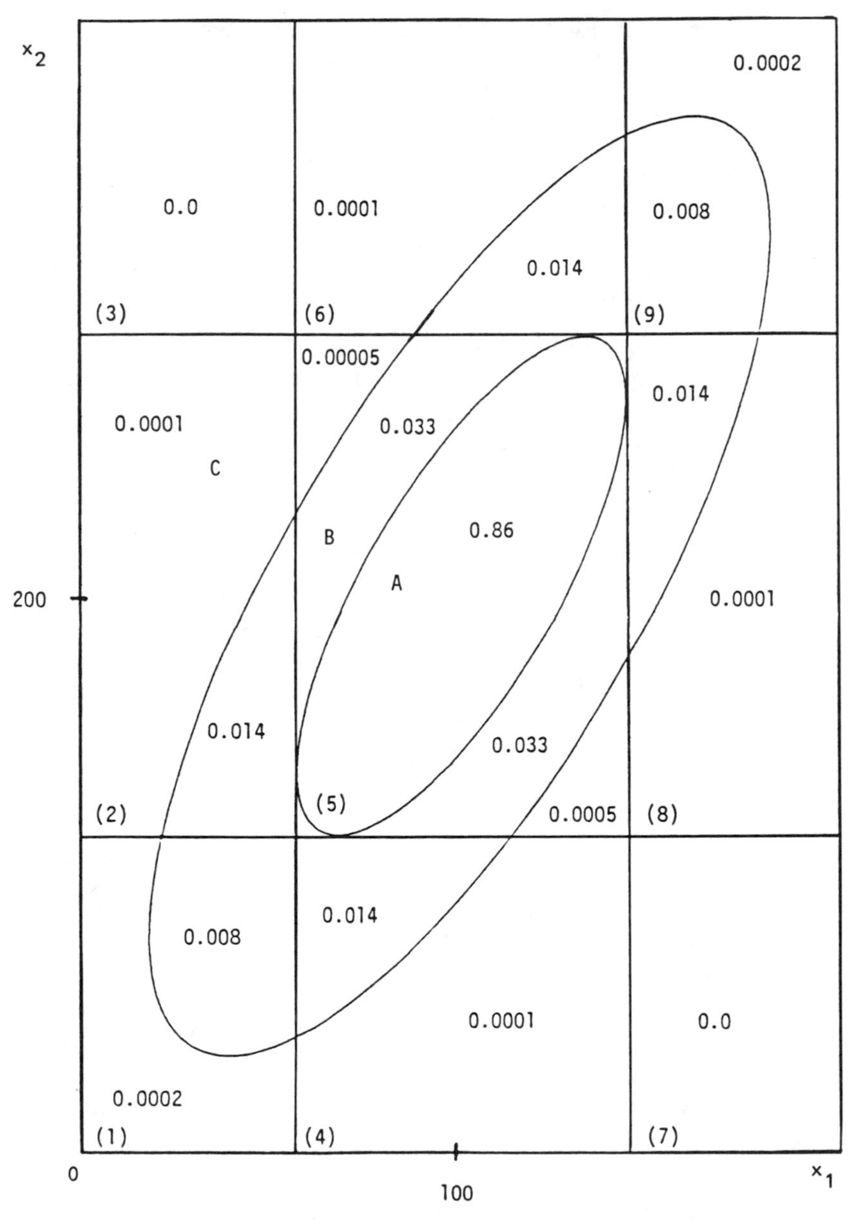

FIGURE 3 Probabilities associated with contours and regions of bivariate normal distribution $\mu = \begin{bmatrix} 100 \\ 200 \end{bmatrix}$ and $\Sigma = \begin{bmatrix} 100 & 112.5 \\ 112.5 & 225 \end{bmatrix}$ with 0.999 and 0.86 probability contours.

TABLE 1 Parameters to Adjust for Sampled Observations in Specified Regions

	1	2	3	4	5	6	7	8	9
1	μ_1 μ_2	μ_1 μ_2	μ_1 σ_{22}	μ_1 μ_2	μ_1 μ_2	μ_1 σ_{22}	μ_2 σ_{11}	μ_2 σ_{11}	σ_{11} σ_{22}
2		μ_1	μ_1 μ_2	μ_1 μ_2	μ_1	μ_1 μ_2	μ_2 σ_{11}	σ_{11}	μ_2 σ_{11}
3			μ_1 μ_2	μ_1 σ_{22}	μ_1 μ_2	μ_1 μ_2	σ_{11} σ_{22}	μ_2 σ_{11}	μ_2 σ_{11}
4				μ_2	μ_2	σ_{22}	μ_1 μ_2	μ_1 μ_2	μ_1 σ_{22}
5					μ_1 μ_2	μ_2	μ_1 μ_2	μ_1	μ_1 μ_2
6						μ_2	μ_1 σ_{22}	μ_1 μ_2	μ_1 μ_2
7							μ_1 μ_2	μ_1 μ_2	μ_1 σ_{22}
8								μ_1	μ_1 μ_2
9									μ_1 μ_2

$n = 2$ rules. These rules provide more opportunities to identify trends in the process. Better rules may be developed as more investigations are made in the bivariate and multivariate domain. However, as the simulation results discussed in the next section indicate, cases have been identified where these simple rules are clearly superior to $n=2$ rules applied separately to each variable.

These rules do not change with the shape of the bivariate normal distribution (determined by $\rho_{x_1x_2}$). Many are not appropriate for the ellipses shown in Figure 3. However, if one imagines that ellipse rotated 90°, some rules that were previously inappropriate become meaningful. In the simulation discussion in the next section of this paper, perfect knowledge of the situation is assumed and all adjustments are made to the nominal values of the mean(s). Practical applications and other suggested methods for adjusting the process are discussed after the simulation results are presented.

1.3 Simulation Results

Computer simulations were employed to test the efficacy of a subset of the decision rules used for adjustment. A process was assumed to produce product at a uniform rate of 100 pieces per hour. The underlying distribution of the control variables of interest was bivariate normal,

$$\mu = \begin{bmatrix} 100 \\ 200 \end{bmatrix} \quad \text{and} \quad \Sigma = \begin{bmatrix} 100 & 112.5 \\ 112.5 & 225 \end{bmatrix}.$$

In the simulations, a steady upward or downward shift in one or both of the means was imposed. The rate of the shift in the means was 1 unit per hour on variable 1 and 2 units per hour on variable 2. Thus, with means of 100 and 200, respectively, and a simultaneous steady upward shift in both means, Item (piece) 1 was assumed to come from a process with $\mu_1 = 100.01$ and $\mu_2 = 200.02$; Item 50 would be representative of a process with $\mu_1 = 100.50$ and $\mu_2 = 201.00$; etc. For each scenario of shifts in mean(s), 50 simulations were employed. One simulation run is concluded when the sampled observation is found to be defective. A sample of size n was drawn from the process in a uniform interval 10 min either side of the hour. Thus, after 1 hr, the sample was randomly selected from

Items 84 through 116, and the corresponding means due to the sustained shifts were in the range $\mu_1 = 100.84–101.16$ and $\mu_2 = 201.68–202.32$. After 2 hr, the sample was selected from Items 184 through 216 and, assuming that the sample items were such that no decision to adjust the mean(s) was made, the sample represented items having means in the range $\mu_1 = 101.84–102.16$ and $\mu_2 = 203.68–204.32$.

Samples of size n were drawn at the times indicated above, and the rules described in the previous section were applied to determine whether an adjustment was to be made or the process was to be unchanged. An adjustment consisted of resetting mean(s) to their nominal values. The shifts were reapplied after each adjustment. Each simulation run ended with the appearance of an observation outside the outer ellipse in the sample.

The goal of the computer simulation study was to determine how well the rules worked in detecting a sustained shift. Of particular interest is an estimate of the fraction defective produced by the process, F_D (where the smaller the fraction, the better the process), and the *average number of items produced between adjustments* (average run length), ARL (where the smaller the value, the more frequent the adjustment). A defective is defined as an item that falls outside of the 0.999 ellipse.

Table 2 presents $n = 2$ and $n = 3$ simulation results for a variety

TABLE 2 Simulation Results Using Bivariate Decision Rules

	$n = 2$		$n = 3$	
	F_D	ARL	F_D	ARL
$\Delta\mu_1\uparrow$	0.010	771	0.007	718
$\Delta\mu_1\downarrow$	0.011	807	0.007	730
$\Delta\mu_2\uparrow$	0.013	580	0.008	532
$\Delta\mu_2\downarrow$	0.014	560	0.010	552
$\Delta\mu_1\uparrow$, $\Delta\mu_2\downarrow$	0.021	332	0.014	280
$\Delta\mu_1\downarrow$, $\Delta\mu_2\uparrow$	0.020	316	0.013	327
$\Delta\mu_1\uparrow$, $\Delta\mu_2\uparrow$	0.012	631	0.007	579

of shifts (Δ) in a positive ($\uparrow$) or negative ($\downarrow$) direction of μ_1 and μ_2. When $n = 2$ and with a shift in either (but not both) means, less than 2% defective product is produced using the proposed adjustment rules. With simultaneous shifts in both means, the worst case ($n = 2$) resulted in 2.1% defective product.

Table 3 compares the results of using bivariate and univariate rules for the case $\Delta\mu_1\uparrow$, $\Delta\mu_2\downarrow$. This case was selected for study because the comparison indicates where the bivariate rules provide a marked improvement over the univariate. Univariate rules for $n = 3, 4, \ldots, 10$ are presented in Table 4. These rules were selected from typical risk-based plans (Ott, 1975) as well as those based on cost (Yu, 1983). As can be seen, when a shift in opposite directions is

TABLE 3 Comparison of Simulation Results Using Bivariate and *Univariate* Rules for $\Delta\mu_1\uparrow$, $\Delta\mu_2\downarrow$

n	F_D	ARL
2	0.021	332
2	*0.086*	*650*
3	0.014	280
3	*0.079*	*692*
4	0.010	342
4	*0.030*	*382*
5	0.009	262
5	*0.078*	*730*
6	0.009	270
6	*0.049*	*619*
7	0.007	255
7	*0.022*	*359*
8	0.005	239
8	*0.045*	*550*
9	0.005	242
9	*0.033*	*528*
10	0.005	217
10	*0.040*	*507*

TABLE 4 Univariate Rules for Multiple Sample Size

n	k^a	Maximum yellow for no adjustment[b]
2	1.50	1
3	1.90	1
4	1.65	1
5	2.00	2
6	1.78	2
7	1.53	2
8	1.64	3
9	1.75	3
10	1.80	3

[a]Indicates the number of units of standard deviation on both sides of the mean considered to be in the green region.
[b]Indicates the number of yellow observations allowed in the sample before a adjustment is indicated.

applied to the means ($\Delta\mu_1\uparrow, \Delta\mu_2\downarrow$), the bivariate rules yield superior results.

Furthermore, when a univariate assumption is made about components of a bivariate vector, the actual risk levels differ considerably from the computed. In the case of $n = 2$, the assumed alpha level associated with making an adjustment according to the univariate rules in the mean for one component is 0.04. For our sample data, the true probability of making an adjustment on one component when there has been no process change is 0.017. For our bivariate rules applied to bivariate data, when $n = 2$, the alpha level is actually 0.02. A comparison of beta risks is a far more complicated issue and is left for future research.

1.4 Future Directions

Some of the popularity of a narrow-limit gaging has been due to its intuitively simple structure. Extended to the multivariate case it loses some of its simple structure; however, it still retains its conceptual appeal. The foregoing discussion and sample rules illustrate how a multivariate scheme might detect an out-of-control state, identify the problem variable, and signal necessary adjustment. Determining the magnitude of adjustment when the actual is not known was not addressed here, though one might consider the use of principal components.

Another approach would be to do a full multivariate analysis of the probability levels of the observations (i.e., do not map the continuous data to attribute characteristics, within, between, or outside specified controur levels). A multivariate extension of the method proposed by Coleman (this volume, Chapter 9) may be a worthwhile direction.

2. A STOCHASTIC CONTROL FRAMEWORK

A fundamental concept underlying the classical Shewhart control charts has been the assumption that the quality characteristics associated with a sequence of items are mutually independent Bernoulli random variables with a constant probability associated with manufacturing a defective part. This assumption is frequently not satisfied in an industrial situation. In these cases, Duncan's (1956) proposed construct, in which the probability of producing a good or bad quality product is dependent on whether the process is in an in-control or out-of-control state, seems more realistic. Furthermore, if in an out-of-control state, the process would remain there until corrective action could be taken.

Stochastic control is concerned with modeling and controlling a dynamic process that changes over time according to some probabilistic laws. The univariate stochastic control methods discussed by Box and Jenkins (1976) are well known and widely applied in continuous process industries. Multivariate stochastic control methods have been developed but have not been widely applied due to the amount of effort required by the modeler (e.g., MacGregor and Wong, 1980).

The Kalman filter is an adaptive stochastic control method that has been extensively studied and applied to both univariate and multivariate data (Kalman, 1960). Recently, it has been studied for monitoring tool wear (Takata et al., 1985) and univariate quality characteristics (Chapter 7, this volume).

In this section, an extension of the Kalman filter, the Harrison–Stevens model (Harrison and Stevens, 1971), is introduced for SPC applications. Also called a multiprocess dynamic linear model, this model incorporates both the notions of state and the adaptive features of the Kalman filter recursive model. The objective is to introduce the concepts and framework of the model and to suggest how it might be formulated for automated SPC.

2.1 Model Definition

The Harrison–Stevens forecasting model predicts the value of an observation based on the dynamics of the underlying parameter set driving the system. This prediction is accomplished by the incorporation of two dynamic equations, one governing the system parameters and the other the observed values. The notation of Bolstad (1986) is used to describe the model.

Let S_t be the state random variable at time t, where each observation of S_t is considered independent and coming from a multinomial distribution. Let

$$\text{Prob}(S_t = j) = \pi_t(j) \qquad \text{for} \qquad j = 1, \ldots, N \tag{5}$$

be the probability of being in state j at time t, given that there are N possible states. Then, given that the system is in state j, $S_t = j$, the parameter equation is

$$\theta_t = A_t'\theta_{t-1} + v_t \tag{6}$$

where θ_t is the $p \times 1$ parameter vector at time t; the $p \times p$ coefficient matrix A_t is assumed to be known at time t; and v_t is the $p \times 1$ parameter perturbation vector, assumed to be normally distributed with zero mean vector and known $p \times p$ variance–covariance matrix $V_t(j)$. This matrix is both time- and state-dependent.

The equation used to forecast the univariate observation, y_t, is given as

$$y_t = B_t'\theta_t + \varepsilon_t \tag{7}$$

where B_t' is the transpose of the $p \times 1$ coefficient vector, assumed to be known at time t and independent of the observation errors ε_t. These errors are assumed independently, identically distributed $N(0, \sigma_\varepsilon^2)$.

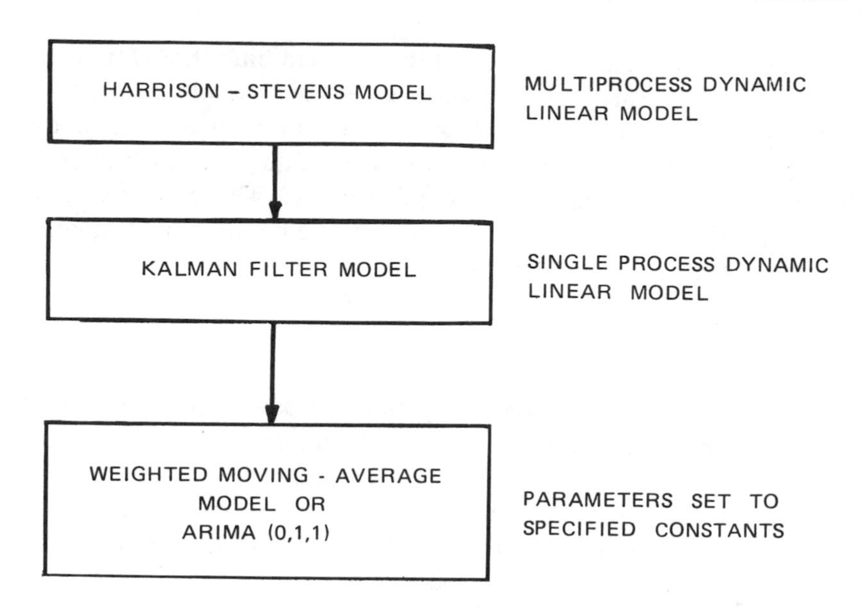

FIGURE 4 Generality of Harrison–Stevens model.

In the case when there is only one state, $N = 1$, the Harrison–Stevens model reduces to the single dynamic linear model called the Kalman filter model. Furthermore, it has been shown that when $p = 1$, $A_t = 1$, $B_t = 1$, and $V_t(j) = \sigma_v^2$ in Eqs. (6) and (7), the Kalman filter model is equivalent to the exponentially weighted moving average model (Kirkendall, this volume, Chapter 6; Abraham and Ledolter, 1986). In addition, the weighted moving average model is the same as an ARIMA(0, 1, 1) model with the parameter greater than zero. Hence, the Harrison–Stevens model represents the most general case. This is illustrated in Figure 4.

2.2 SPC Application

The multiprocess model differs from the linear model by the incorporation of the notion of states. Depending on the state of the system and time of the observation, the parameter vector θ_t of the model is subject to perturbations. The concept is that, under some conditions (or states), the underlying model (the parameter model) will experience some disturbances (perturbations), causing the observation (y_t) to fluctuate. Applying this model to a SPC environment enables the processing system state to be classified as being in control or out of control.

The Harrison–Stevens model has been traditionally used for short-term forecasting when the discontinuities or changes in the state of the underlying system are known (Johnston and Harrison, 1980). It has also been used, however, to identify changes in state (Smith et al., 1983). In the field of process control, both uses must be addressed. For instance, the start-up of a process represents a series of known changes in the system. The Harrison–Stevens model may be useful for allowing for these changes, yet still being able to detect marked deviations in state. Once the processing system has been operating in a state of statistical control, this model may then become useful for identifying when the system suddenly goes out of control.

One of the major benefits of using the multiprocess dynamic model in the SPC framework is the general structure that it encompasses. For purposes of illustration, consider that the system may be in one of two states, in control ($j = 1$) or out of control ($j = 2$). The in-control state implies that the observation y_t is varying about some constant mean value. The out-of-control state suggests that the underlying parameter values have changed. For example, let the parameter vector be defined as

$$\theta_t = \begin{bmatrix} \mu_t \\ \omega_t \\ \delta_t \end{bmatrix} \quad \begin{matrix} \text{mean}_t \\ \text{slope}_t \\ \text{transient}_t \end{matrix} \tag{8}$$

Figure 5(a) shows a plot of observed values for a process in an in-control state. Figure 5(b–d) illustrates how the observed values may indicate a change in any one of the components; i.e., the process is out-of-control.

Consider the coefficient matrix for the parameter dynamics suggested by Bolstad (1986):

$$A_t = \begin{bmatrix} 1 & 1 & 0 \\ 0 & 1 & 0 \\ 0 & 0 & 0 \end{bmatrix} \quad \text{and} \quad B_t = \begin{bmatrix} 1 \\ 1 \\ 1 \end{bmatrix} \tag{9}$$

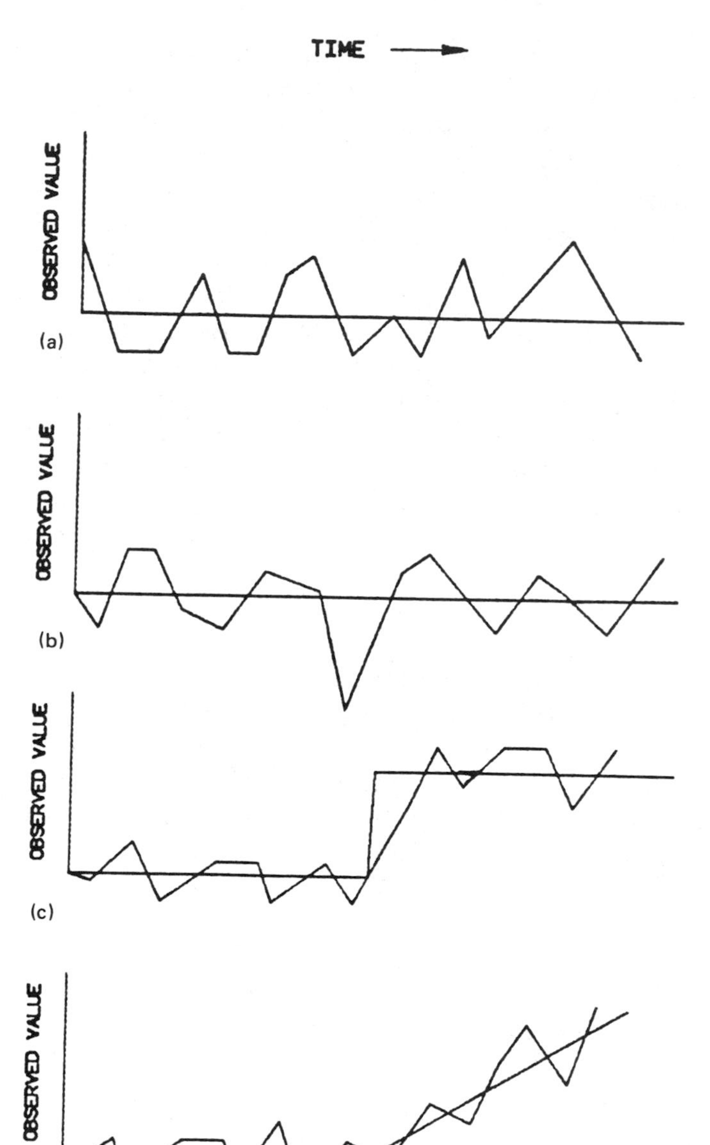

FIGURE 5 Possible process states: (a) in control, (b) transient out of control, (c) change to different process level (out of control), and (d) trend to out of control.

Then

$$\theta_t = \begin{bmatrix} \mu_t \\ \omega_t \\ \delta_t \end{bmatrix} = \begin{bmatrix} \mu_{t-1} + \omega_{t-1} \\ \omega_{t-1} \\ 0 \end{bmatrix} + v_t \tag{10}$$

and let

$$V_t(1) = \begin{bmatrix} 0 & 0 & 0 \\ 0 & 0 & 0 \\ 0 & 0 & 0 \end{bmatrix} \quad \text{and} \quad V_t(2) = \begin{bmatrix} \sigma_\mu^2 & 0 & 0 \\ 0 & \sigma_\omega^2 & 0 \\ 0 & 0 & \sigma_\delta^2 \end{bmatrix} \tag{11}$$

If the system is in a state of statistical control at time $t - 1$ and remains there at time t, then $\omega_{t-1} = 0$ and $\mu_t = \mu_{t-1} = \mu$; that is, $\theta_t' = [\mu, 0, 0]$ and $y_t = \mu + \varepsilon_t$. However, if the system goes out of control at time t, then the parameter vector is subject to perturbations at time t,

$$\theta_t = \begin{bmatrix} \mu_{t-1} \\ 0 \\ 0 \end{bmatrix} + v_t \sim N[0, V_t(2)] \tag{12}$$

and

$$y_t = \mu_t + \omega_t + \delta_t + \varepsilon_t \tag{13}$$

Consequently, if the system goes out of control, the observed values are driven by a mean, a slope, or a transient component.

These parameters are not observable and must be estimated, given the observed values, y_t. Invoking some preliminary assumptions about the behavior of the parameter estimates (the usual normality assumptions), Harrison and Stevens proposed a recursive algorithm for estimating the parameter values. These estimates are then used in the observation equation to enable forecasts to be made. In applying this model to statistical process control, the goal is to use the recursive algorithm for making real-time forecasts to detect an out-of-control state prior to observing in y_t severe departures from the mean. These forecast errors may then be used to construct a feedback loop in the system, which would be useful for self-adapting behavior.

The recursive algorithm for estimating the parameters at time t was given by Harrison and Stevens (1971, 1976). It is based on the recursive result of the Kalman filter equation, which is stated below. [Derivation of the result may also be found in Crowder (this volume, Chapter 7).]

Given the Kalman filter equations

$$\theta_t = A_t\theta_{t-1} + v_t \tag{14}$$

$$y_t = B_t'\theta_t + \varepsilon_t \tag{15}$$

where θ_t is a $p \times 1$ parameter vector, A_t is the $p \times p$ parameter coefficient matrix, B_t' is the transpose of the $p \times 1$ observation coefficient matrix, and $v_t \sim (0, V)$ and $\varepsilon_t \sim N(0, \sigma_\varepsilon^2)$. Define $\mathbf{y}_{t-1} = (y_{t-1}, y_{t-2}, \ldots, y_0)$, i.e., all the observations up to and including that at time $t-1$. Then

$$\hat{\theta}_{t-1} = E(\theta_{t-1}|\mathbf{y}_{t-1}) \tag{16}$$

and

$$e_t = y_t - B_t'\hat{\theta}_{t-1} \tag{17}$$

and assume $(\theta_{t-1} | \mathbf{y}_{t-1}) \sim N(\hat{\theta}_{t-1}, \Sigma_{t-1})$ with Σ_{t-1} known. Then

$$(\theta_t | \mathbf{y}_t) \sim N(A_t\hat{\theta}_{t-1} + C_{12}(C_{22})^{-1}e_t; \Sigma_t) \tag{18}$$

where

$$\Sigma_t = C_{11} - C_{12}C_{22}^{-1}C_{21}$$

and

$$C_{11} = A_t\Sigma_t A_t' + V \qquad C_{12} = C_{11}B_t$$

$$C_{21} = B_t'C_{11} \qquad C_{22} = \sigma_\varepsilon^2 + B_t'C_{11}B_t \tag{19}$$

Furthermore,

$$(e_t | \mathbf{y}_{t-1}) \sim N(0, C_{22}) \tag{20}$$

The term $C_{22} = B_t'C_{11}B_t + \sigma_\varepsilon^2$ is the variance matrix associated with the "one-step-ahead" forecast made at time $t-1$. The product $C_{12}(C_{22})^{-1}$ is called the Kalman gain matrix. The coefficients of the conditional distribution represent the recursive update equations for the Kalman filter equations.

Bolstad (1986) gives a thorough description of the derivation of the multiprocess recursive algorithm. The interested reader is referred to that article, of which only the highlights are presented here.

Recall the model specified in Eqs. (5)–(7) and define

$$\text{Prob}(S_{t-1} = i|\mathbf{y}_{t-1}) = q_{t-1}(i) \tag{21}$$

Under the assumption that the random variables of the states are independent in time,

$$\text{Prob}(S_t = j, S_{t-1} = i) = \Pi_t(j)q_{t-1}(i) \tag{22}$$

Let

$$(\hat{\theta}_{t-1}|S_{t-1} = i, \mathbf{y}_{t-1}) \sim N(\theta_{t-1}, R_{t-1}(i)) \tag{23}$$

where $R_{t-1}(i)$ is assumed known given the $S_{t-1} = i$ and all past observations up to y_{t-1}. Then

$$\begin{aligned} & f(\hat{\theta}_{t-1}, i|\mathbf{y}_{t-1}) \\ & \quad = q_t(i)(2\pi)^{-p/2}|R_{t-1}(i)|^{-1/2} \\ & \qquad \times \exp[-\tfrac{1}{2}(\hat{\theta}_{t-1} - \theta_{t-1})'[R_{t-1}(i)]^{-1}(\hat{\theta}_{t-1} - \theta_{t-1})] \\ & \qquad \text{for} \quad i = 1, \ldots, N \end{aligned} \tag{24}$$

Define

$$W_t(i,j) = A_t R_{t-1}(i)A_t' + V_t(j) \tag{25}$$

where $V_t(j)$ is as described above for Eq. (6). The predicted parameter vector at time $t + 1$ is a mixture random vector

$$(\hat{\theta}_{t+1}|\mathbf{y}_t) = A_{t+1}(\hat{\theta}_t|y_t) = \sum_{i=1}^{N} q_t(i)A_{t+1}(\hat{\theta}_t|S_t = i, y_t) \tag{26}$$

Then

$$(\hat{\theta}_{t+1}|y_t) \sim N\left[\theta_{t+1}; \sum_{j=1}^{N} \sum_{i=1}^{N} q_t(i)\Pi_{t+1}(j)W_{t+1}(i,j)\right] \tag{27}$$

This equation is used to obtain the forecasted value

$$\hat{y}_{t+1} = B_{t+1}'(\hat{\theta}_{t+1}|y_t) \tag{28}$$

which leads to a prediction error, $e_t = y_{t+1} - \hat{y}_{t+1}$ and

$$e_t \sim N(0, \sum\sum q_t(i)\Pi_{t+1}(j)B_{t+1}'W_{t+1}(i,j)B_{t+1} + \sigma_\varepsilon^2) \tag{29}$$

Therefore, this variance matrix is associated with the "one-step-ahead" forecast made at time t, and is analogous to that in Eq. (20).

The probability parameters $q_t(i)$ and $\Pi_{t+1}(j)$ act as weights for the mean vector $(\hat{\theta}_t|\mathbf{y}_t)$ and $(\hat{\theta}_{t+1}|\mathbf{y}_t)$. These may be estimated using the current prediction error; that is, at time $t-1$ after y_{t-1} is observed and again after y_t. These probabilities may themselves be used to identify changes in the state of the system.

2.3 Concluding Comments

The intent of this discussion has been to introduce a general framework for the use of the dynamic linear model in statistical process control. Alwan and Roberts (this volume, Chapter 4) have shown that the ARIMA(0, 1, 1) model is useful when dealing with correlated data in statistical process control. This Harrison–Stevens model is a more general case, and this preliminary investigation indicates that the future research opportunities in this area are rich.

ACKNOWLEDGMENTS

The author wishes to thank J. B. Keats for his suggestions and M.-C. Chua for his programming assistance.

REFERENCES

Abraham, B., and J. Ledolter (1986). Forecast functions implied by autoregressive integrated moving average models and other related forecast procedures. *International Statistical Review, 54,* 51–66.

Alt, F. B. (1985). Multivariate quality control. In *Encyclopedia of Statistical Sciences,* Vol. 6, eds. S. Kotz and N. L. Johnson, pp. 110–122. Wiley, New York.

Bolstad, W. M. (1986). Harrison–Stevens forecasting and the multiprocess dynamic linear model. *The American Statistician, 40,* 129–135.

Duncan, A. J. (1956). The economic design of $\bar{X}$ charts used to maintain current control of a process. *Journal of the American Statistical Society*, *51*, 228–242.

Harrison, P. J., and C. F. Stevens. (1971). A bayesian approach to short-term forecasting. *Operational Research Quarterly*, *22*, 341–362.

Harrison, P. J., and C. F. Stevens. (1976). Bayesian forecasting. *Journal of the Royal Statistical Society B*, *38*, 205–228.

Jackson, J. E. (1956). Quality control methods for two related variables. *Industrial Quality Control*, *12*, 4–8.

Jackson, J. E. (1959). Quality control methods for several related variables. *Technometrics*, *1*, 359–377.

Johnston, F. R., and P. J. Harrison. (1980). An application of forecasting in the alcoholic drinks industry. *Journal of the Operational Research Society*, *31*, 699–709.

Shewhart, W. A. (1931). *Economical Control of Quality of Manufactured Products*. Van Nostrand Reinhold, Princeton, N.J.

Smith, A. F. M., M. West, K. Gordon, M. Knapp, and I. C. Trimble. (1983). Monitoring kidney transplant patients. *Statistician*, *32*, 46–54.

Takata, S., M. Ogawa, P. Bertok, J. Ootsuka, K. Matushima, and T. Sata. (1985). Real-time monitoring system of tool breakage using Kalman filtering. *Robotics and Computer Integrated Manufacturing*, *2*, 33–40.

Walker, P. G. (1985). Specification, comparison and control of inter-related multivariate product properties. *Quality Assurance*, *11*, 67–70.

SECTION III

Innovative Techniques for Statistical Process Control

9
Generalized Control Charting

David E. Coleman*
RCA David Sarnoff Research Center
Princeton, New Jersey

1. INTRODUCTION

An alternative to the traditional $\bar{X}$ control chart can be developed by making some simple transformations of raw data. As with *cumulative sum* (CUSUM), this alternative has the advantages of arbitrarily good approximation to individual confidence levels, shift detection and trend detection without ad hoc supplementary runs rules, and data-driven assessment of the duration of shifts and trends. Its performance appears to be comparable to that of CUSUM. Unlike CUSUM, the approach described in this paper can be used for process variables of arbitrary distribution.

This *generalized control charting* (GCC) approach can be briefly described as follows:

1. Characterize the parent distribution of a process variable by its estimated *cumulative distribution function* (cdf). As an option,

---

**Current affiliation:* Alcoa Technical Center, Alcoa Center, Pennsylvania.

one can use subgroup means of m observations (m sufficiently large), and apply the *central limit theorem* (CLT) to give (approximately) the gaussian cdf (Φ).

2. Apply the probability integral transform, using the cdf of (1). This produces percentage points that are approximately uniformly distributed on (0, 1). Analogous to the Shewhart chart, any one value with a percentage point as low as, say, 0.001 or as high as, say, 0.999 would arouse suspicion that the process mean has shifted low or high.

3. If we assume the values are independent, the product of the successive percentage points is an indicator of how low any set of sequential values is relative to the parent distribution. (Similarly, the product of one minus the percentage points is an indicator of how high any set of sequential values is.) p-values for the product are derived from the quantity $-2 \log(\text{product})$, which has the χ^2 distribution—using the technique commonly called "the combination of independent tests of significance" (Johnson and Kotz, 1970, pp. 63–64).

4. By plotting raw data plus graphic indicators of the unlikelihood (1 − likelihood) of having observed recent sequences of values, one can dynamically choose an operating confidence level, and can let the p-values computed from the data indicate the location and the length of intervals of unusually low or high data. Also, if the process is stable, we can in principle collect enough historical data to set approximation error as low as desired.

2. THE GENERALIZED CONTROL CHARTING METHOD

We assume independent random variables X_i, with common parent density f, and with IID outcomes x_i, $i = 1, \ldots, n$. Density f is unknown, but can be approximated, possibly using the CLT and the gaussian cdf if we choose to use subgroup means as the x_i. Let F be the cdf for f, with approximation $\hat{F}$. Then $U_1 = F(X_1)$, $U_2 = F(X_2), \ldots, U_n = F(X_n)$ are uniformly distributed on (0,1), with outcomes we will call μ_i.

An obvious but ill-conceived strategy is to use the product of successive μ_i terms to monitor a process, say,

$$w_{n,k} := \prod_{i=n-k+1}^{n} u_i$$

This is the probability of seeing k successive values this low, or lower on a point-wise basis:

$$w_{n,k} = \Pr[X_{n-k+1} \leqslant x_{n-k+1}, X_{n-k+2} \leqslant x_{n-k+2}, \ldots, X_n \leqslant x_n]$$

which is the hyper-volume of the k-dimensional rectangular parallelepiped with a vertex at the origin, and other vertices determined by the $\{u_i\}$ on the edges of the k-dimensional unit hypercube. This is *not* what we want, for at least two reasons. First, $w_{n,k}$ is order-independent, so it does not take advantage of the order information (though we could examine successive $w_{n,k}$ terms for trends, such as $\{w_{n+j,k+j}\}$ for $j \geqslant 0$). Second, and more serious, is a problem of interpretation and significance. A sequence, $x_{n-k+1}, x_{n-k+2}, \ldots, x_n$, each approximately at the median, would have $w_{n,k} \approx (\frac{1}{2})^k$. This is a very small number for moderately large values of k, but such a series should *not* be declared "out of control: too low." Also, the equivalent of $w_{n,k}$ on the "high side" is

$$w'_{n,k} := \prod_{i=n-k+1}^{n} (1 - u_i) = \Pr[X_{n-k+1} \geqslant x_{n-k+1}, \ldots, X_n \geqslant x_n]$$

which for the same sequence of x_i terms would have a $w'_{n,k} \approx (\frac{1}{2})^k$. Hence, a sequence of values at about the median level could be simultaneously declared "out of control: too low," and "out of control: too high." This is not reasonable.

Instead, consider the *distribution* of $\prod_{i=n-k+1}^{n} U_i$. If we define $Z_i := -2\log(U_i)$, then Z_i is distributed as $p(t) = \frac{1}{2}e^{-t/2}$, which is the χ_2^2 distribution. Note that the sum of the Z_i,

$$\sum_{i=n-k+1}^{n} Z_i = -2\left[\sum_{i=n-k+1}^{n} \log(U_i)\right] = -2\log\left(\prod_{i=n-k+1}^{n} U_i\right)$$

is distributed as a χ_{2k}^2. This is because of the self-reproducing property of a χ^2 distribution (Johnson and Kotz, 1970, Vol. 2, pp. 63–64 and Vol. 1, 193–194); the sum of χ^2 *random variables* (r.v.'s) is itself a χ^2 r.v. with degrees of freedom equal to the sum of the degrees of freedom of its components. That is,

$$\left(\sum_{i=1}^{k} \chi^2_{d_i}\right) \quad \text{is distributed as} \quad \chi^2_{\left(\sum_{i=1}^{k} d_i\right)}$$

If we know the distribution of $-2\log(\cdot)$, we can get the distribution of its argument, since $-2\log(\cdot)$ is strictly monotonic. Specifically, for any random variable $Y_{n,k}$ and any constant $r > 0$, $\Pr[Y_{n,k} < r] = \Pr[-2\log(Y_{n,k}) > -2\log(r)]$, for $Y_{n,k} > 0$. For this application, we set

$$Y_{n,k} := \prod_{i=n-k+1}^{n} U_i$$

with outcomes

$$y_{n,k} = \prod_{i=n-k+1}^{n} u_i$$

Note that $Y_{n,k}$ has the same distribution as does $Y_k := \prod_{i=1}^{k} U_i$.

How do we interpret $p_{n,k}^{\text{low}}$, defined as $p_{n,k}^{\text{low}} = \Pr[Y_k < R]$ where $R = y_{n,k}$? Figure 1 shows how we can represent this probability in the case $n = k = 2$. For this case, we have plotted the unit square, which is the joint sample space for U_1 and U_2. The pairs (U_1, U_2) are uniformly distributed over this unit square, since the joint distribution is the product of independent uniforms. Hence, the area of any region in the square represents the probability of an outcome in that region. The value $w_{2,2}$ represents the rectangular area within the unit square defined by the axes and the point (u_1, u_2). This point is on the hyperbola defined by $r = U_1 U_2$, where we set constant $r := u_1 u_2$. For any x_1 and x_2, we get $u_1 = F(x_1)$, and $u_2 = F(x_2)$, and $r = U_1 U_2$ defines a truncated hyperbola of equiprobable outcomes. It provides the upper right boundary to a region $(r \leqslant U_1 U_2)$, the area of which, $p_{2,2}^{\text{low}}$, is the probability of observing a sequence of two values this low, *or rarer in the sense of being low.* This probability is the χ_2^2 cumulative function of $-2\log(r)$, which can be found in tables, or computed using standard formulas (Abramowitz and Stegun, 1972).

The pairs of values that fall within a region described by low r are pairs where at the extremes: (a) either x_1 or x_2 is extraordinarily low, and the other is not unusual, or (b) x_1 and x_2 are low as a pair.

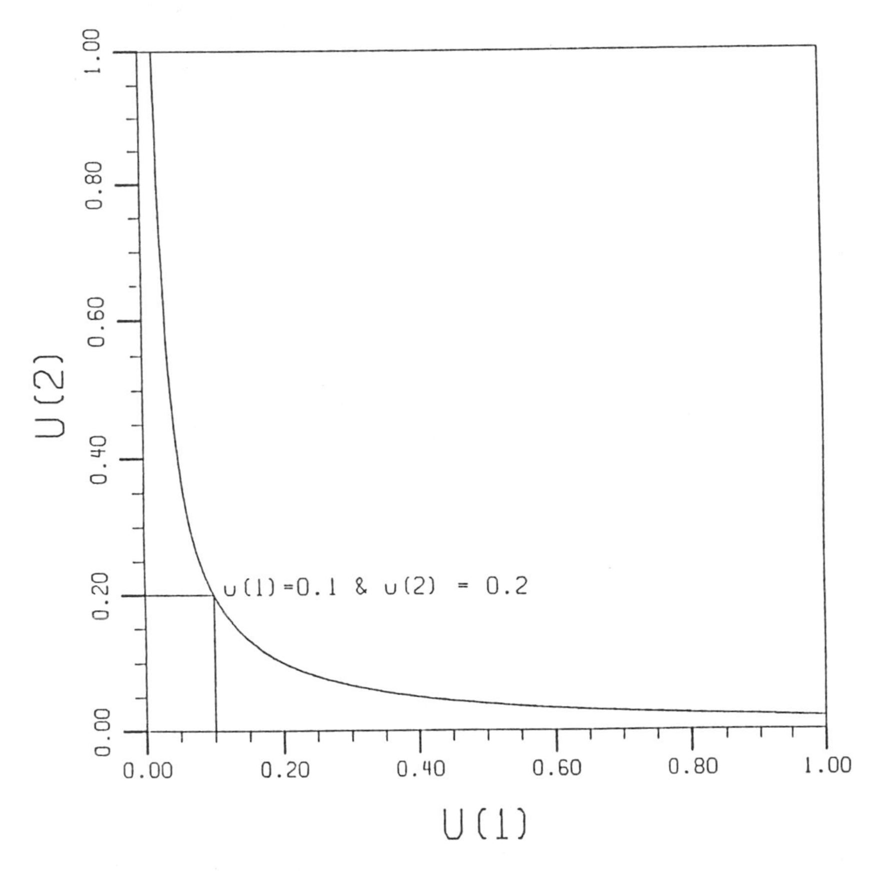

FIGURE 1 Sequence of two $X(i)$ values, product $u(1)u(2) < 0.02$.

Note that this potentially allows more sensitivity than does CUSUM control charting when shift, S, from target T is less than the specified CUSUM shift parameter K, and when there is decreasing variability; the GCC method will signal a shift given a stream of values $T + S, T + S, \ldots$, while a CUSUM will not. This scenario is quite common in *automatic test equipment* (ATE), which can get "stuck" and send out the same, erroneous measurement for every unit tested. This behavior is more thoroughly evaluated through a comparison of run-length distributions in Section 4.

The GCC procedure is generalized to higher dimensions, which correspond to longer sequences of process variable values. The

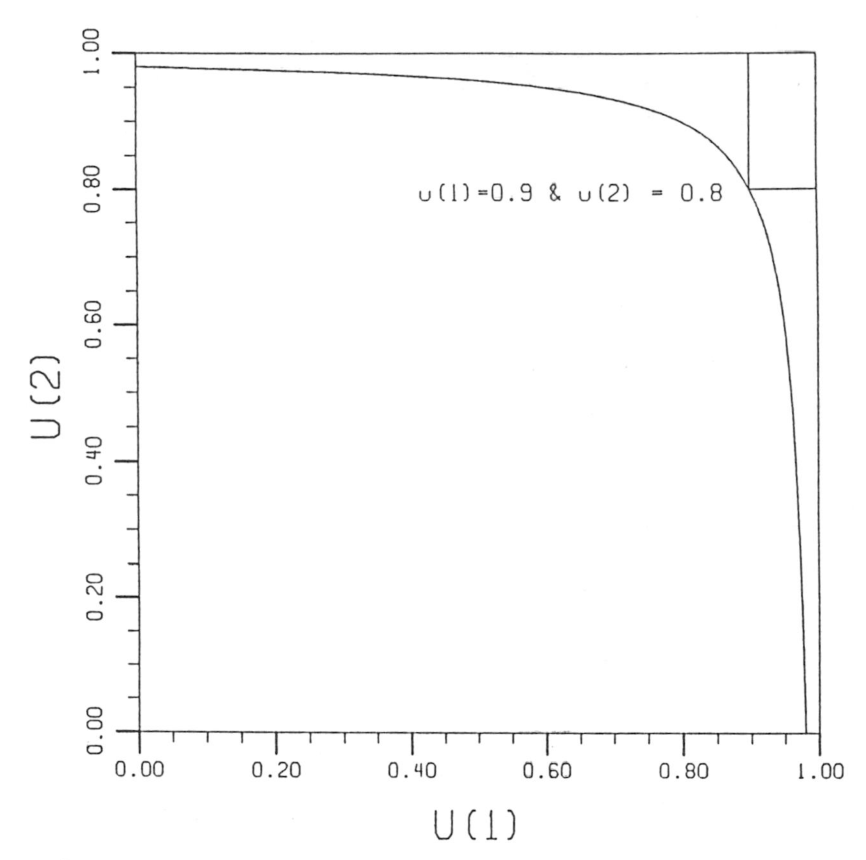

FIGURE 2 Sequence of two $X(i)$ values, product $[1 - u(1)][1 - u(2)] < 0.02$.

$k = 2$ unit square (for any k) generalizes to a unit hypercube, within which the hypervolume represents probability, since probability is uniformly distributed throughout the hypercube. The truncated hyperbola becomes a truncated hyperboloid. We compute cumulative probabilities defined by the k-dimensional hypercube axes and the truncated hyperboloid by consulting chi-squared tables or formulas for $2k$ degrees of freedom, for sequence length k. The probability has the same type of interpretation as for the $k = 2$ case. A low value of the probability $p_{n,k}^{\text{low}}$ is interpreted to indicate that the sequence of k values is unusually low. That is, the likelihood of seeing k values in a row that are this low or rarer—in the sense of being low—is $p_{n,k}$. This does not preclude there being some high values in the sequence of k values. It may be the case that some contiguous subset of l of the k values is even *more* improbably low.

We take the same approach, in symmetry, for high values as we do for low values, as seen in Figure 2. Here we define $v_i := 1.0 - u_i$, and analogously compute p-values $p_{n,k}^{\text{high}}$.

What about interval length? How far back into history should we go? The GCC algorithm allows us to identify unusual sequences of arbitrary length, but that capability is of limited practical use. A priori, we must choose a maximum interval length, $k_{\max}$, over which we will seek for trends or shifts. We also probably want to look at all contiguous subsets for even more unusual sequences. Simultaneous inference is an issue here, because we will be looking at such a vast number of percentage points. The α-level, $p_{\min}$, against which we compare values of p^{low} and p^{high} depends on $k_{\max}$. The greater the maximum interval length, the smaller must be the nominal α-level to achieve the same actual Type I error rate [and the same *average run length* (ARL), though run-length distributions may vary among $\{k_{\max}, p_{\min}\}$ pairs with the same error rate or ARL].

3. METHODS OF PRESENTATION

To illustrate the use of the GCC procedure, with different methods of presentation, synthetic datasets were generated in Minitab on a Prime 750 computer. These data sets are shown in Figures 3–7. The data sets are all synthesized $N(0, 1)$ data that were modified to exhibit various deviations of interest in SPC: a shift in the mean, a transient value, a trend, cyclical behavior, and an increase in variation.

3.1 GCC Glyph Plot

An unorthodox but informative plot for presenting $p_{n,k}$ values from the GCC algorithm is the glyph plot. The glyph used in Figure 3(b) is a "check mark" glyph [see the enlarged Figure 3(b)]. The check mark shape was selected to make it easy to compare any $p_{n,k}$ value

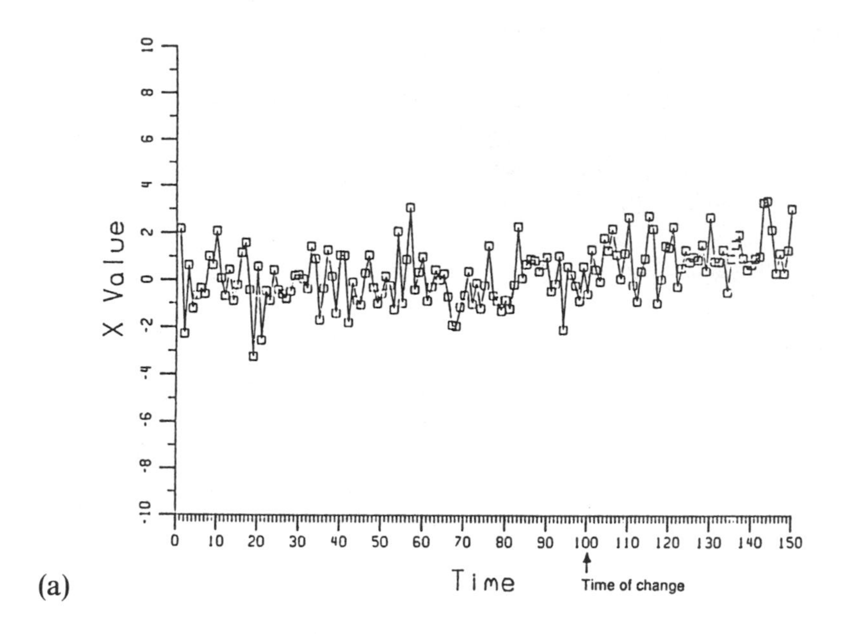

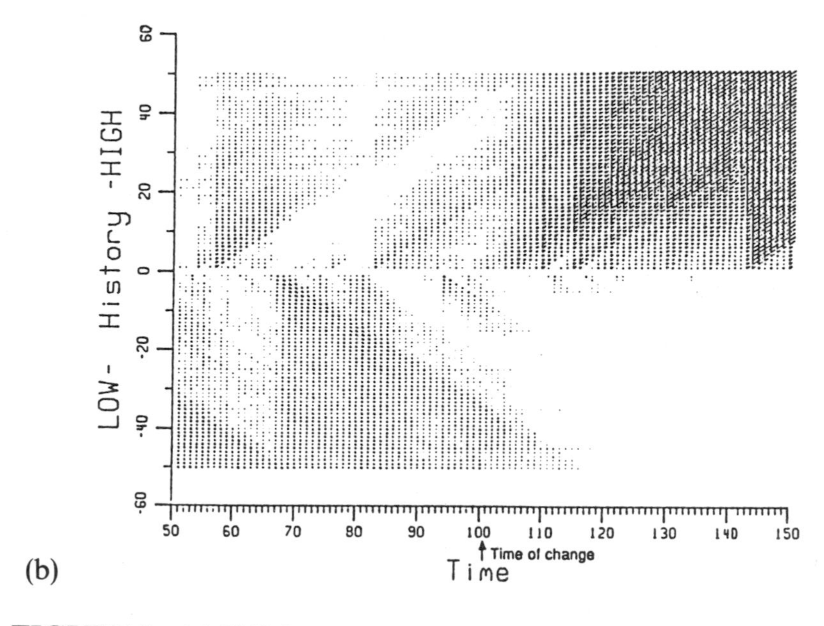

FIGURE 3 (a) Shift by σ at time = 100. (b) GCC glyph plot of shift by σ. (c) GCC three-dimensional plot of shift by σ.

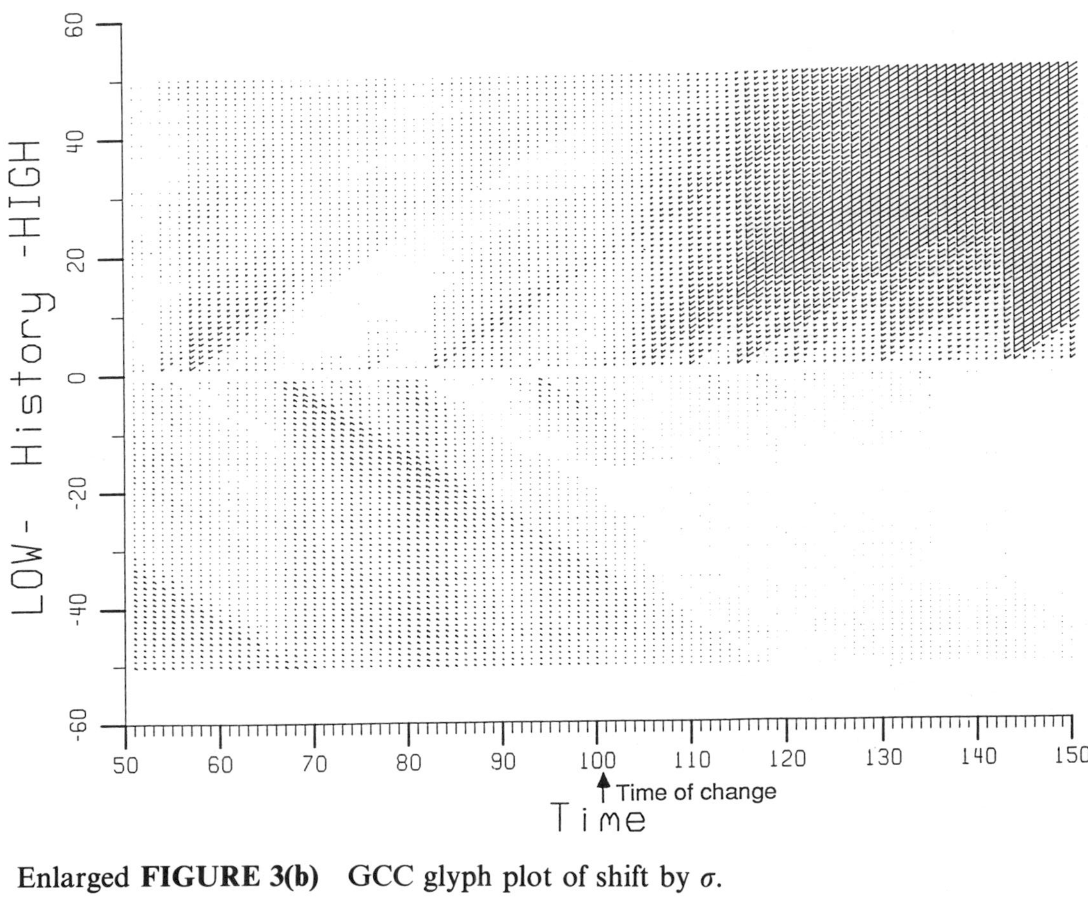

Enlarged **FIGURE 3(b)** GCC glyph plot of shift by σ.

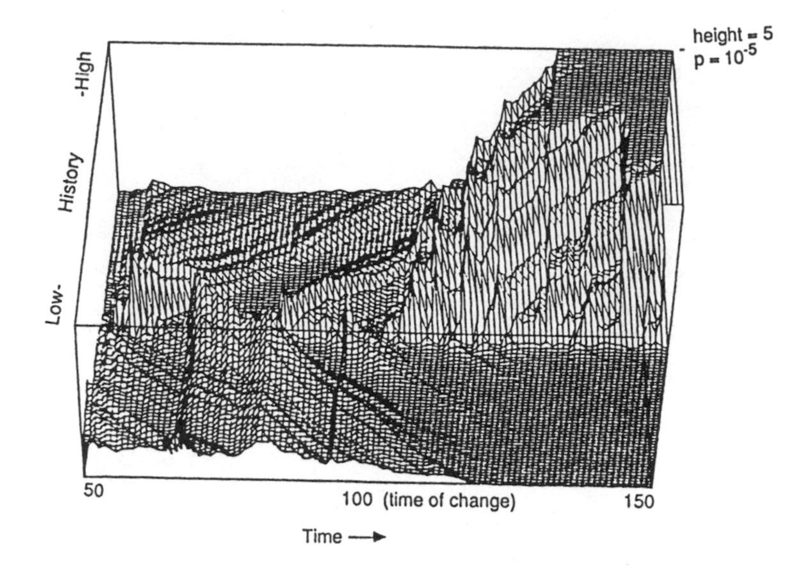

(c)

FIGURE 3 (continued).

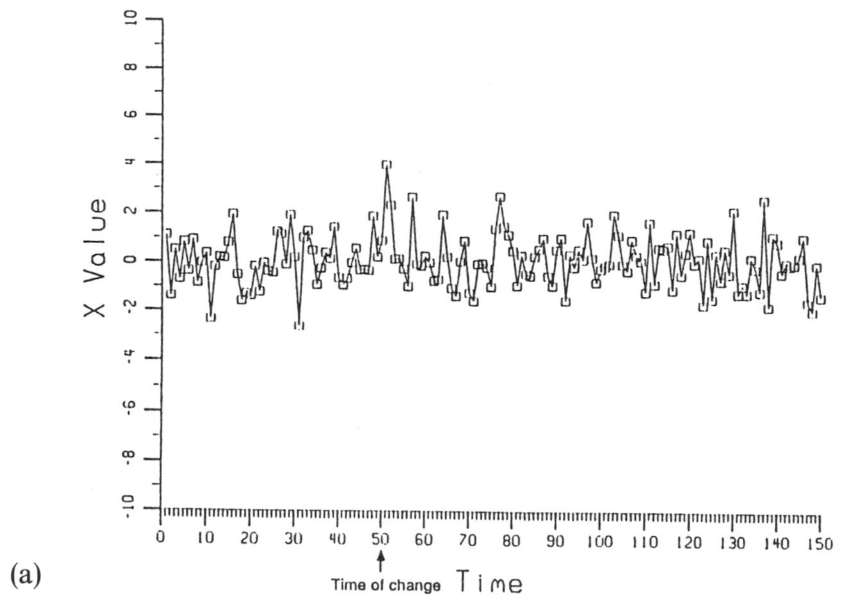

(a)

FIGURE 4 (a) Transient value at time = 50, $x(50) = 4.0$. (b) GCC glyph plot of transient value. (c) GCC three-dimensional plot of transient value.

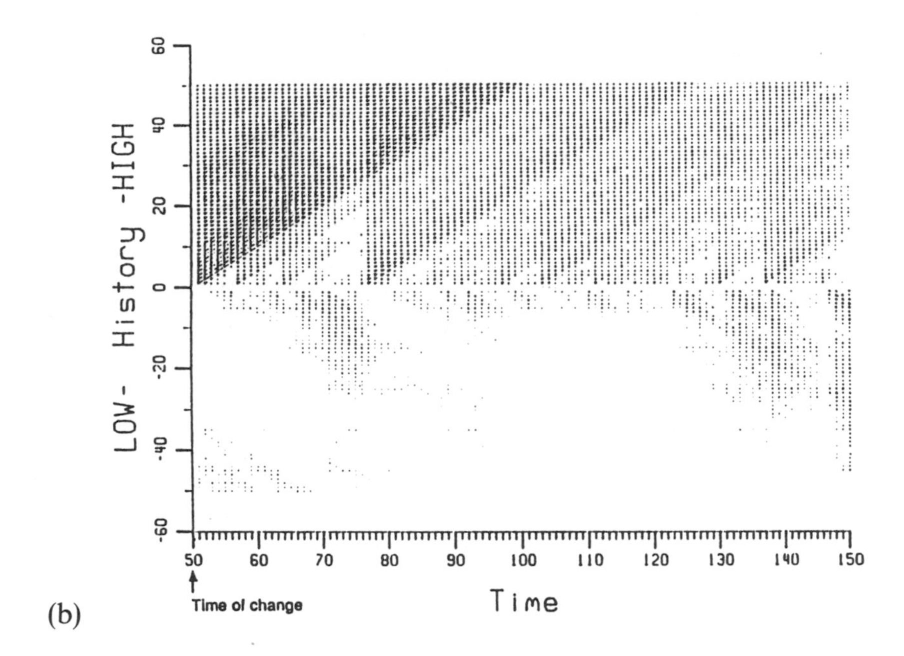
LOW- History -HIGH
60
40
20
0
-20
-40
-60
50
60
70
80
90
100
110
120
130
140
150
Time of change
Time

(b)

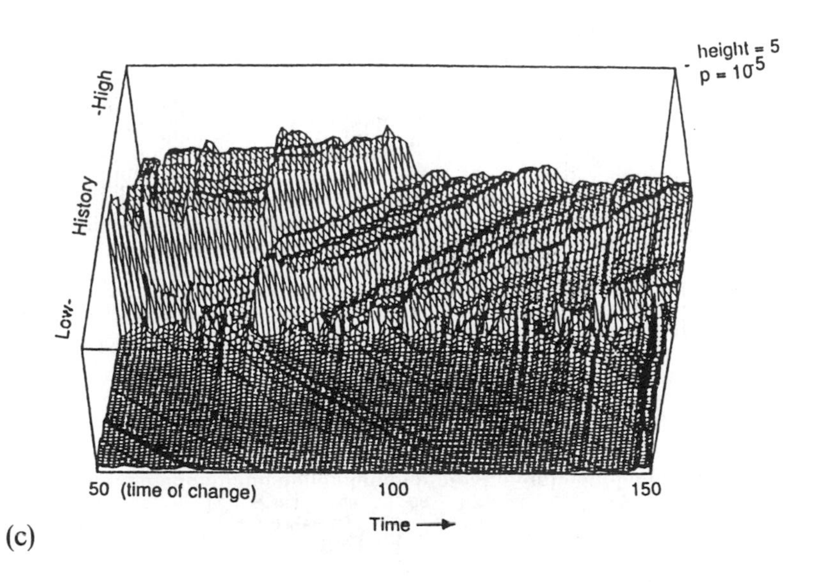
height = 5
p = 10⁻⁵
-High
History
Low-
50 (time of change)
100
150
Time

(c)

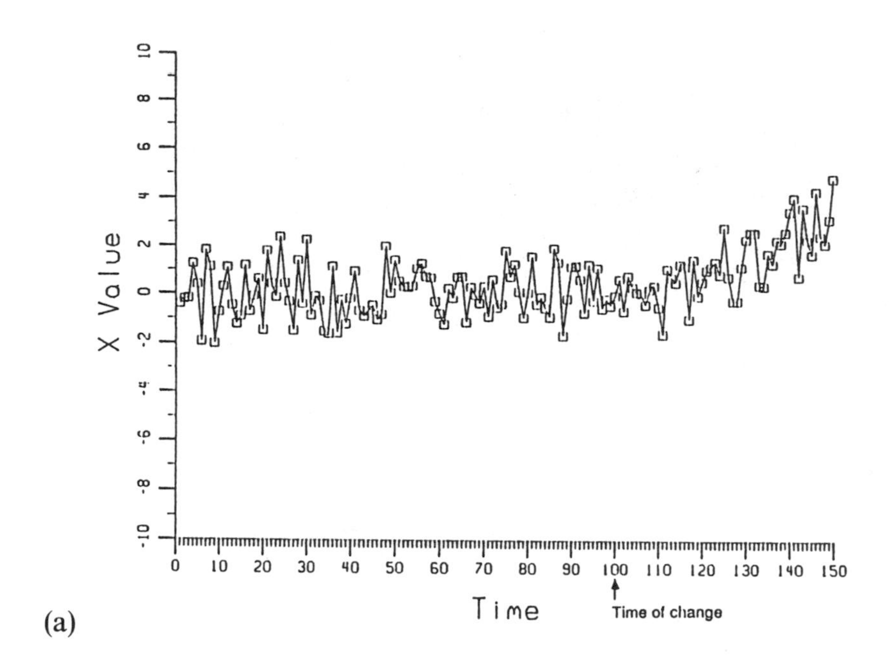

(a)

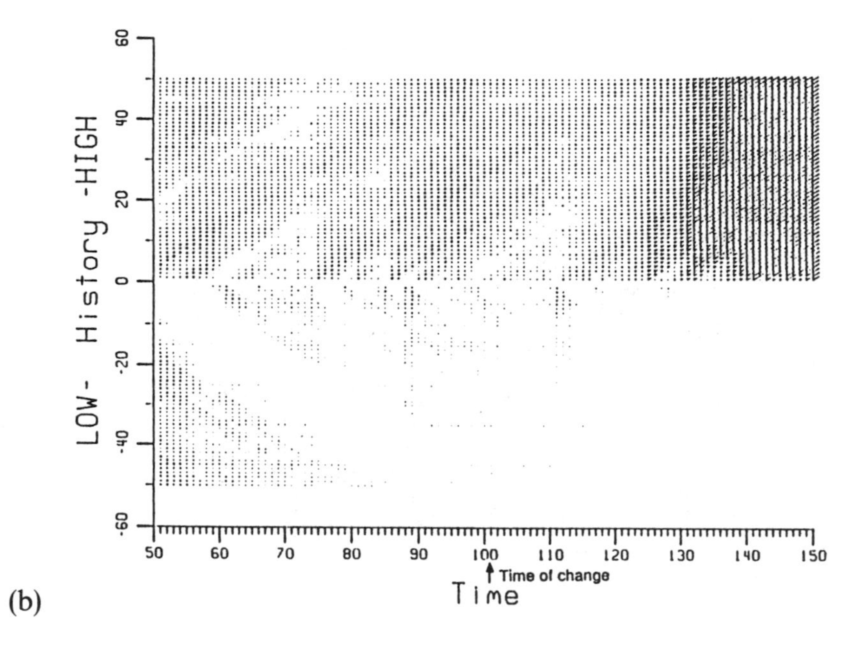

(b)

FIGURE 5 (a) Trend of 0.05σ per subgroup at time = 100. (b) GCC glyph plot of trend. (c) GCC three-dimensional plot of trend.

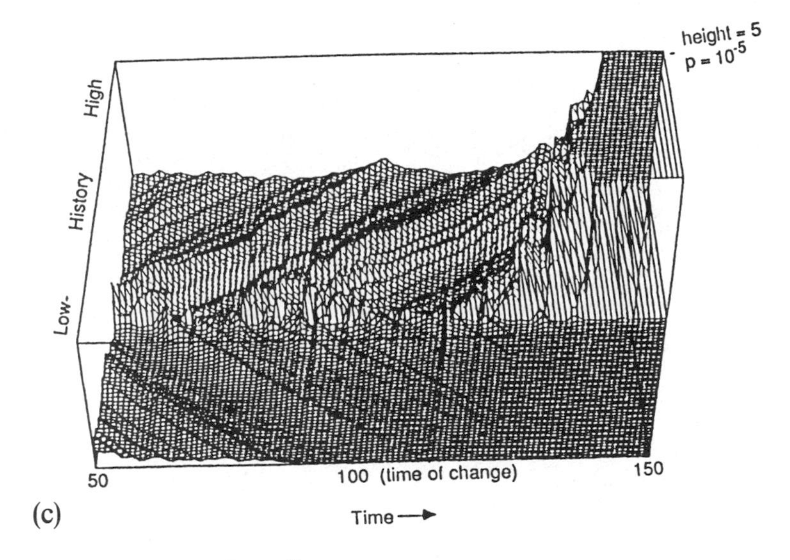

(c)

FIGURE 5 (continued).

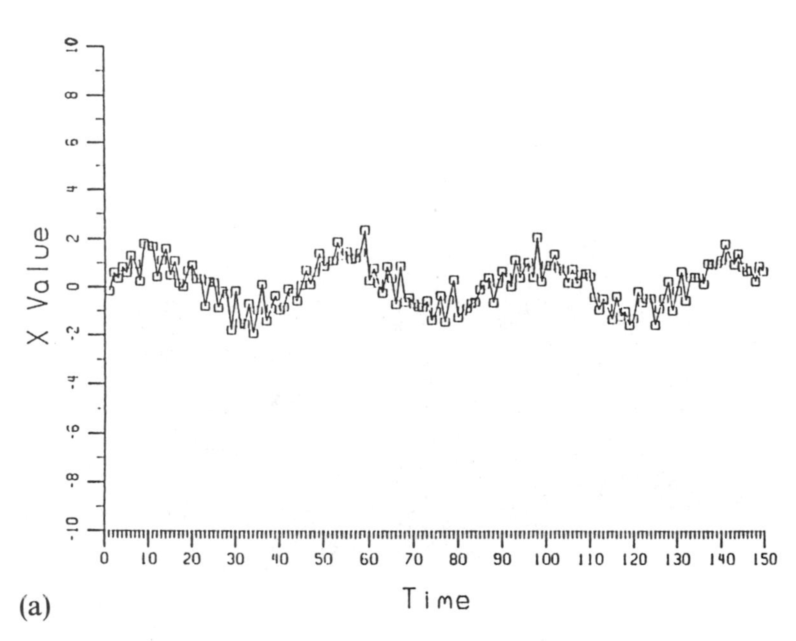

(a)

FIGURE 6 (a) Cyclical data. (b) GCC glyph plot of cyclical data. (c) GCC three-dimensional plot of cyclical data.

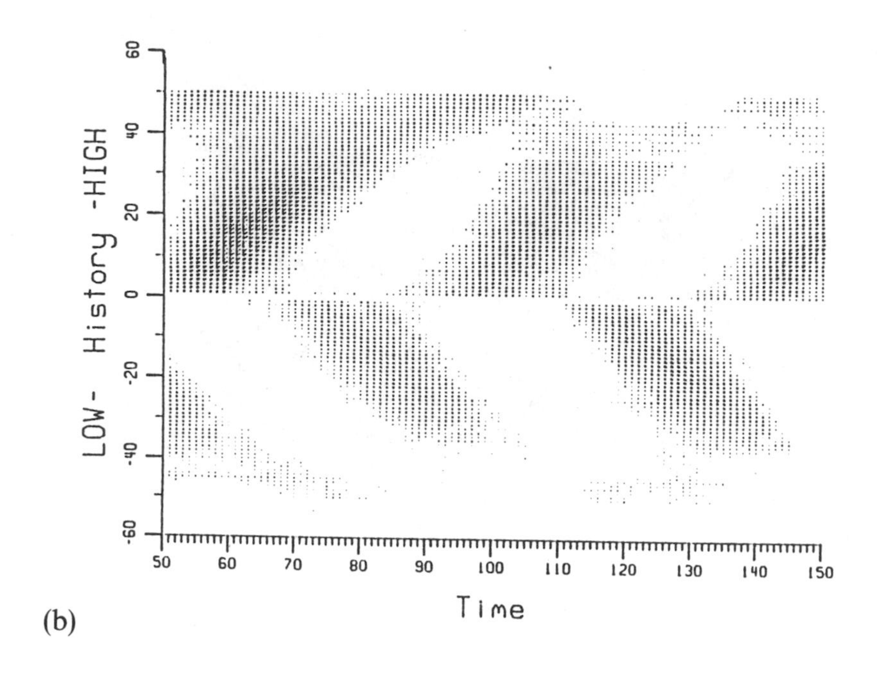

(b)

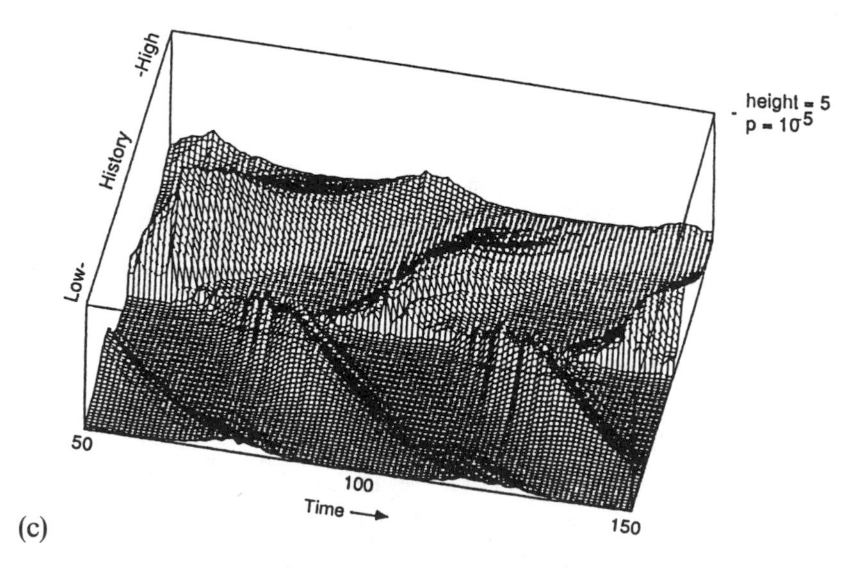

(c)

FIGURE 6 (continued).

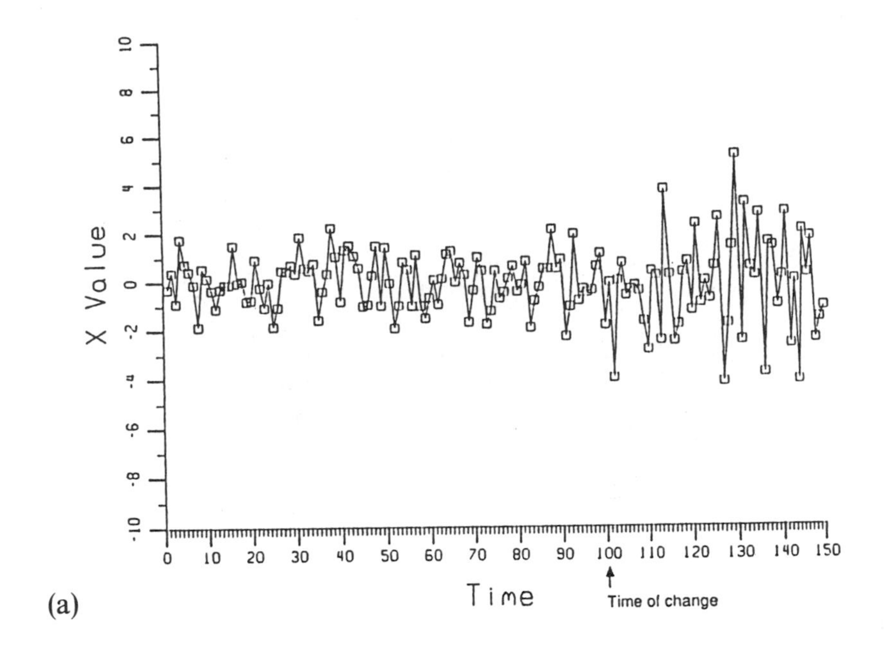

(a)

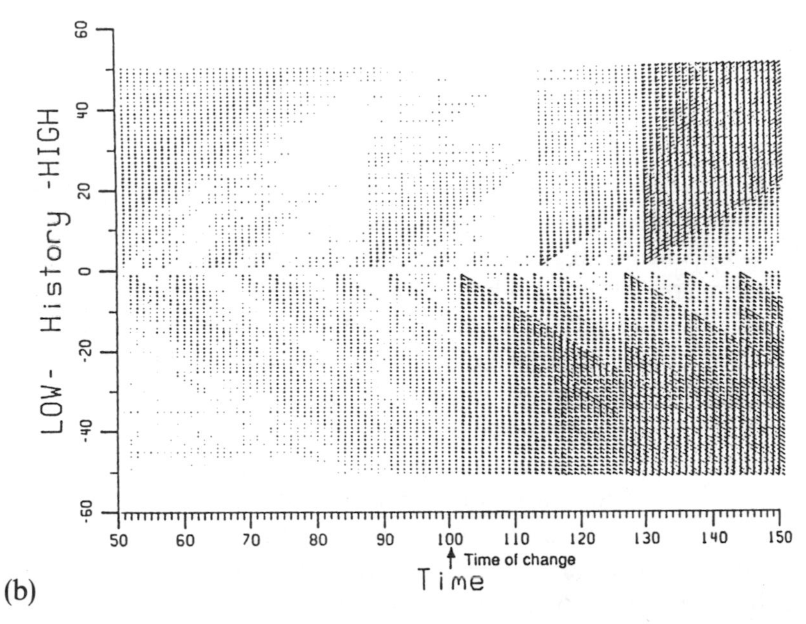

(b)

FIGURE 7 (a) Heteroscedasticity ($\sigma \rightarrow 2\sigma$) at time = 100. (b) GCC glyph plot of heteroscedasticity. (c) GCC three-dimensional plot of heteroscedasticity.

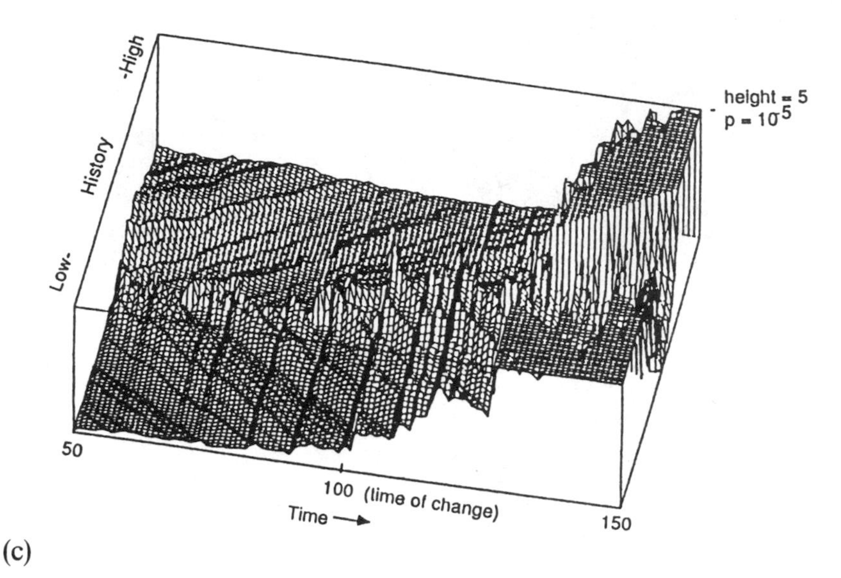

(c)

FIGURE 7 (continued).

with its neighboring values, vertically, and along $\pm 45^\circ$ lines to the right (that is, we compare $p_{n,k}$ to $p_{n,m}$ and $p_{n+j,k+j}$, where $k > m > 0$, and $j > 0$). The glyph grows in size as the probability shrinks, with a maximum size attained when its extending line segments collide with neighboring glyphs. In a glyph plot, such as Figure 3(b), the horizontal axis is the time axis. More properly, it is the axis for the index i of x_i, as is common in standard control charts. The vertical axis is split. In the upper half of the plot, it indicates for how many time increments j we compute high-side $[i, j]$ probability, counting back from time i. Similarly, the vertical axis in the lower half of the plot indicates the low-side $[i, (-j)]$ probability associated with the sequence ending at time i, and going back $j - 1$ time intervals, so as to include j points. Note that since $p_{i,j}^{\text{high}} \neq 1 - p_{i,j}^{\text{low}}$ for all $j > 1$, the upper part of the plot is not simply a negative relief of the lower. To highlight low probabilities, we make glyph size proportional to $q_{i,j} = -\log_{10}(p_{i,j})$, suitably scaled, and perhaps truncated at some maximum value. (We may choose, for example, to put a ceiling of 5 on every value of $q_{i,j}$, corresponding to a minimum displayed probability of 10^{-5}.) A large glyph at $[i, j]$ indicates that the probability is low—under the null hypothesis of a stationary process—of seeing a sequence of i points as high or less likely (in the sense of being high) as was seen ending at point j. A large glyph at $[i, -j]$ has the analogous interpretation.

We can make these observations of the chart—given by construction: (1) the p values bordering the central horizontal axis are simply $\hat{F}(x_i)$ on the upper side, and $1 - \hat{F}(x_i)$ on the lower side, so they give the same point-wise information as an x chart or simple Shewhart $\bar{X}$ chart; (2) the p values successively further from the horizontal axis, in a direction parallel to the vertical axis (j increasing in magnitude, i constant) indicate to what extent sequences ending at time interval i confirm what is seen with point x_i; (3) the p values along a $\pm 45°$ line from $p_{i,1}$ to the right and away from the horizontal axis show to what extent points following x_i confirm what was seen with the point x_i; (4) conversely, the $\pm 45°$ line of p values from $p_{i,j}, j > 1$ to the left and towards the horizontal axis show to what extent the behavior shown by x_{i-j+1} is confirmed by subsequent points.

From these observations, we can derive some simple heuristics:

1. A shift in the mean [Figure 3(a)] will appear as a region of generally growing glyphs tending toward the upper right or lower right of the plot, while the complementary part of the plot becomes flat [Figure 3(b)].
2. A transient value, either very high [Figure 4(a)] or very low, will appear as a large glyph near the horizontal axis, with a region of decreasing glyphs trailing out along a $\pm 45°$ line to the right. Neither the values that precede or follow the transient will "confirm" that a permanent shift has taken place [Figure 4(b)].
3. A trend [Figure 5(a)] will appear similar to a shift, but there will be an overall acceleration in glyph size [Figure 5(b)].
4. Cyclical behavior [Figure 6(a)] will result in undulating glyph regions, possibly on the high side and low side—in which case a "fish-bone-like" figure results [Figure 6(b)].
5. An increase in variation [Figure 7(a)] will result in many isolated large glyphs along the horizontal axis, eventually

leading to simultaneous out-of-control flags for moderate to long sequences on both the low and high sides. This is a result of points appearing in the high-dimensional analog of the overlap regions determined by the low-side and high-side hyperbolas, such as could be seen by overlapping Figures 1 and 2. This may be close in appearance to a high rate of transient shifts [Figure 7(b)].

3.2 GCC Three-Dimensional Plot

A dramatic alternative to the glyph plot is a three-dimensional "mountain plot," such as seen in Figure 3(c). This plot requires no loss of the information contained in the glyph plot, but may be harder to present and interpret. Instead of encoding $-\log_{10}(p_{n,k})$ with glyph size, we encode it with height. To ensure a reasonable scaling, we round up very low probabilities to 10^{-5} as was done for glyph plots. Figures 3(c), 4(c), 5(c), 6(c), and 7(c) correspond to Figures 3(a), 4(a), 5(a), 6(a), and 7(a) (they were generated with the same data).

3.3 GCC Strip Plot

For those who prefer a linear plot, the GCC probabilities can be condensed, as shown in Figure 8. Above and below the center strip—where the data are plotted—symbols indicate the minimum probability seen at that time index, n, selected from some set of probabilities that include that index. Choices for that set include: (a) points $p_{n,m}$ on the vertical line at time n, for any m; (b) points on the $\pm 45°$ line to the right, $p_{n+j,j+1}$; (c) points contained within the triangular region bounded by and including the lines in (a) and (b), which is all probabilities for sequences which include x_n. Only choice (a) allows us to incrementally update the GCC strip plot. Choices (b) and (c) require us to revise earlier values in an ex post facto manner, which is often not useful. Choice (a) was used to generate Figure 8.

The data plotted in the center strip can be the original values, x_i, or transformed, u_i. The former are easier to relate to, while the latter are more convenient because the scale is fixed.

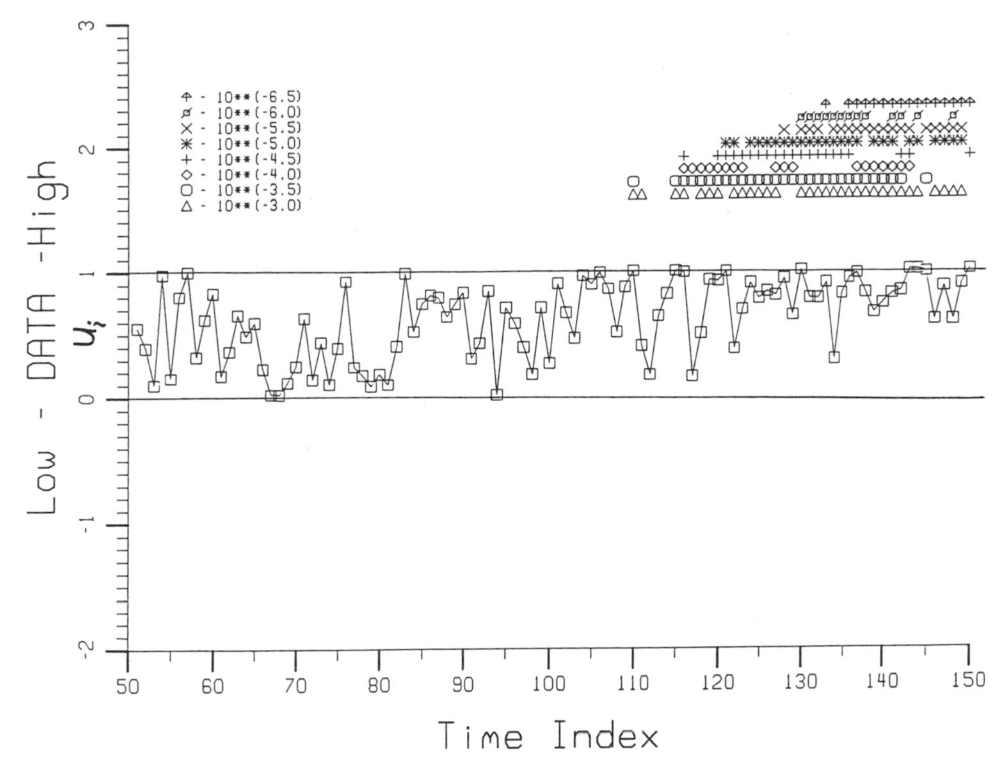

FIGURE 8 GCC strip plot of shift by σ.

4. EVALUATION

4.1 Introduction

Evaluation of a control chart methodology can be done on at least two levels: quantitative, and qualitative. At the very least, the quantitative evaluation should precede any proposed use of a control chart scheme. Of interest is the *operating characteristic curve*, or the distribution of run lengths under various alternative hypotheses. More commonly, a simple comparison of *average run lengths* (ARLs) is made. Since run-length distributions are typically strongly skewed away from zero, and the arithmetic mean is not robust, the ARL is potentially misleading. Unfortunately, its prevalence in the literature makes it hard to avoid using. More appropriate would be run-length distribution quantile summaries, including the median, though comparing quantile levels can become cumbersome.

Qualitative evaluation refers to the degree of success a control chart methodology achieves "in the field." This is affected by many factors: data acquisition, user-friendliness, skill level of user, display or output device, etc. We will not concern ourselves here with these important factors—because they are not inherent to the underlying methodology. What we *are* interested in is the ease of interpretation—both basic interpretation and sophisticated interpretation. How easy is it for a person of little skill or of considerable skill, using the algorithm, to identify nonstationarity? How easy is it for him to keep from being fooled by normal variation? This type of evaluation is necessarily subjective, but certain control charts are clearly easier to interpret than others.

4.2 Monte Carlo Strategy

Quantitative evaluation of the GCC algorithm can be carried out analytically, or empirically—via Monte Carlo experimentation. The former is in progress and will be reported on at a later date. Results of the latter are reported here.

A battery of straightforward Monte Carlo runs was performed to evaluate the GCC algorithm. The author decided to compare GCC to the control chart method best known for good power in the face of small to moderate shifts and trends: CUSUM. CUSUM has been designed to have ARLs superior to those of traditional $\bar{X}$ charts. Its use has been described widely in recent years (Goel and Wu, 1971; Lucas, 1973; Woodall, 1983). However, it has not achieved anywhere near the breadth of use of the $\bar{X}$ chart, as is evidenced by available software (*Quality Magazine*, 1986). This may be due to difficulty of interpretation—a qualitative issue that will be discussed in Section 4.3. The many variants of CUSUM (e.g., CUSUM with Fast Initial Response, and Shewhart/CUSUM) were not considered for comparison, but some published results of evaluation can be found elsewhere (Lucas, 1982; Lucas and Crosier, 1982).

Since CUSUM was selected for comparison, the author selected alternative hypotheses from those for which there are published data on the performance of CUSUM. These included shifts by 0.25, 0.50, 0.75, 1.0, 1.5, 2.0, 2.5, and 3.0 times the standard deviation of the subgroup mean, and trends at rates of 0.001, 0.002, 0.005, 0.01, 0.02, 0.005, 0.1, and 0.2 standard deviations of the subgroup mean per time unit. The null hypothesis case, process variable stationarity, was also studied. Unfortunately, only ARLs could be found published for CUSUM for all of these scenarios (Bissell, 1984, 1985; Asbagh, 1985; Wadsworth et al., 1986).

Two major questions particular to the GCC algorithm were addressed by the Monte Carlo runs:

1. What is a reasonable value p_{min} of $p_{n,k}$ to select, below which we declare "out of control"?
2. What do we gain by computing $p_{n,k}$ for large k, i.e., what is the marginal benefit of stepping a little further back into history?

To partially answer the first question, we first recall that we have a situation of simultaneous inference in computing the many $p_{n,k}$ values. For example, if our maximum interval length, k_{max}, is 50, then after 100 elapsed time units the GCC algorithm will have computed 100·50·2 = 10,000 probabilities (the factor of 2 is because we compute both $p_{n,k}^{low}$ and $p_{n,k}^{high}$). Since these values are highly correlated, however, we need not use a Bonferroni-type $p_{min} = \alpha/10{,}000$. In the Monte Carlo runs, a range of p_{min} levels was explored: 10^{-2}, 10^{-3}, 10^{-4}, 10^{-5}, 10^{-6}. The region most comparable to CUSUM performance (in-control ARL comparable to the CUSUM with $h = 5$, $k = 0.5$) appeared to be between 10^{-3} and 10^{-4}, so this region was explored in increments of $10^{-0.2}$. It should be noted that CUSUM procedures also involve simultaneous inference on sequences of varying lengths (at any given time, the V-mask can be crossed by any number of historical points of any vintage), so an appropriate method of comparison is run-length statistics, *not* nominal error rates.

To partially address the second question, the Monte Carlo software was designed to record out-of-control declarations for various maximum interval lengths: $k_{max} = 5$, 20, and 50. There is, of course, an interaction between the appropriate levels of k_{max} and p_{min} to achieve a given desired ARL profile. It is the opinion of the author that there is little point in selecting a small k_{max} for reasons other than performance. The compute-intensive nature of the GCC method is of minor concern given the advent of fast 16-bit and 32-bit processors.

The Monte Carlo software was written in FORTRAN 77, and executed on a Prime 750 computer. The IMSL gaussian random number generator, GGNML (SP), was used to generate gaussian data sets of length up to 5000. From 100 to 500 data sets were used in each run (a greater number were used if the standard error of the mean was large). Each gaussian data set was then altered, if appropriate, by a shift or trend, starting at the first time increment. Other deviations from stationarity have been included as options in the software, and still others can be trivially programmed. The data were assumed by the GCC algorithm to be perfect gaussian. This is unrealistic relative to actual use of a control chart method, but it is a reasonable basis for evaluation. It is the typical scenario for control chart evaluation. For each data set, the software recorded the time index of the first one-sided declaration of out-of-control (a value of $p_{n,k} < p_{\min}$), for each of the combinations of $p_{\min}$ and $k_{\max}$. Logically, a declaration for $p'_{\min}$ means a declaration for $p''_{\min} > p'_{\min}$ and a declaration for $k'_{\max}$ means a declaration for $k''_{\max} > k'_{\max}$. However, the converse of the second (as well as the first) does *not* hold. Type II errors were also counted by the software, but these results are not shown in this paper.

The time indices of declaration were then processed and stored in Minitab, segregated by each possible combination of $p_{\min}$ and $k_{\max}$. Stem and leaf plots reveal how severely skewed were the run-length distributions.

4.3 Monte Carlo Results

The detailed results of the Monte Carlo approximations of run length distributions are too lengthy to present in this paper. A summary of the ARL results is presented in tabular form in Appendix C. In this section, a small portion of the ARL results is presented in graphical form. Figure 9 compares ARLs for shifts in the mean of the process variable. Figure 10 makes the same comparison for trends. In both figures, the standard chosen for comparison is the oft-recommended CUSUM with $h = 5$, $k = 0.5$.

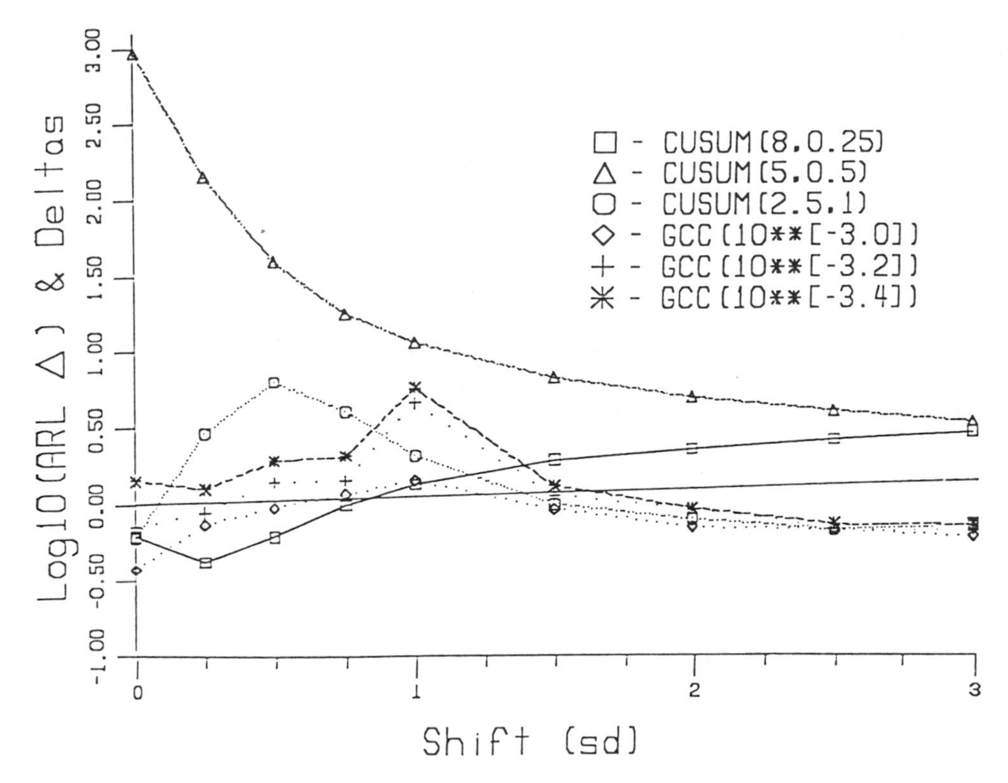

FIGURE 9 Plot of ARL values for three CUSUMs and three GCC schemes, for shifts of different amounts. The CUSUM(5, 0.5) scheme is used as the standard for comparison, represented by the triangle, with $\log_{10}$(ARL) values shown. Other schemes are displayed showing the proportional difference in their ARLs to the ARLs of the standard. A positive value of 0.3 for shift s means an ARL that is 30% greater than the CUSUM(5, 0.5) scheme when the mean has shifted by $s\sigma$.

The $\log_{10}$ values of the ARL for different shift amounts (including 0.0) are indicated by triangles. All of the other symbols show *proportional* deviations from this standard. Two other CUSUMs are shown. They are commonly recommended as alternatives to the standard when a higher sensitivity to very small shifts or slow trends (squares, $h = 8$, $k = 0.25$) or to larger shifts or faster trends

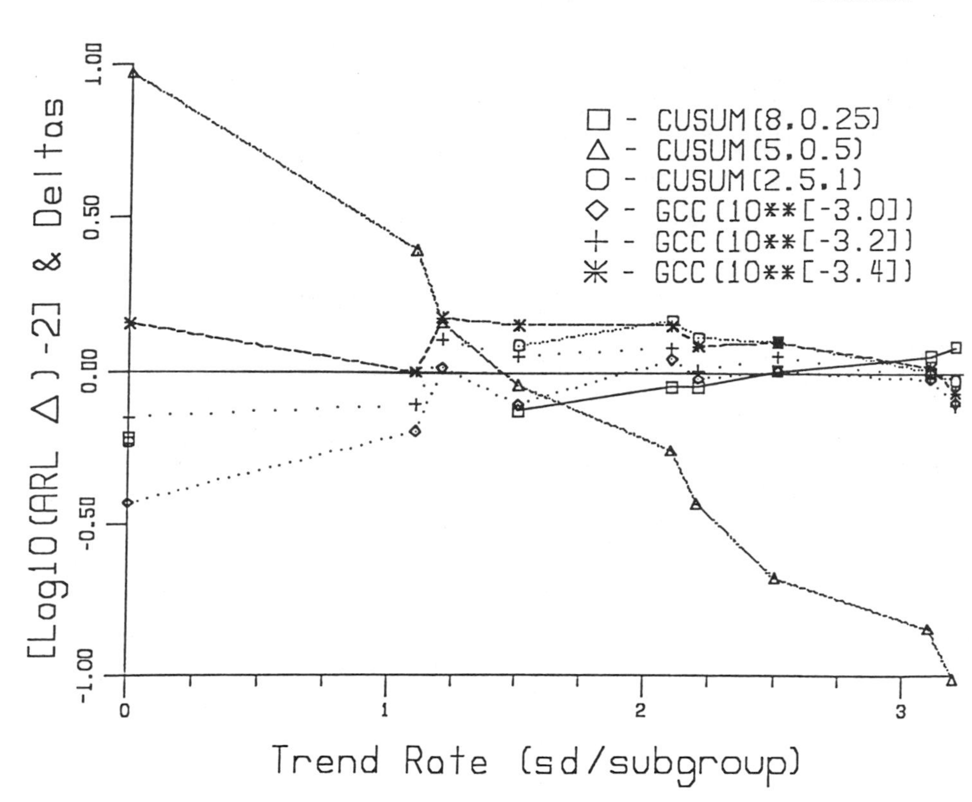

FIGURE 10 Plot of ARL values for three CUSUMs and three GCC schemes, for trends of different amounts. The CUSUM(5, 0.5) scheme is used as the standard for comparison, represented by the triangle, with $\log_{10}(\text{ARL}) - 2$ values shown (the -2 is to make the scale more convenient). Other schemes are displayed showing the proportional difference in their ARLs to the ARLs of the standard. A positive value of 0.3 for shift s means an ARL that is 30% greater than the CUSUM(5, 0.5) scheme when the mean has shifted by $s\sigma$.

(octagons, $h = 2.5$, $k = 1$) is desired. All CUSUM results are from referenced sources (Bissell, 1984, 1985; Asbagh, 1985; Wadsworth et al., 1986). Shown for comparison are the Monte Carlo ARL results for GCC using three different values for $p_{\min}$.

These figures show that the performance of GCC, as measured by ARL, is roughly comparable to that of these CUSUM methods, but does not dominate any of them. In particular, GCC($10^{-3.2}$) is seen to be comparable to CUSUM(2.5, 1). A seemingly worthwhile area of further evaluation of GCC is using $p_{\min}$ values of about $10^{-3.3}$, which should result in a stationary-process ARL that is comparable to CUSUM(5, 0.5). This work is ongoing.

Note that an unfortunate limiting factor in comparison of ARLs is the uncertainty in the estimates, particularly for CUSUM ARLs. The figures show that the CUSUM results are at least smoothly varying, and in the manner that one would expect. The GCC shows what looks like an anomalous blip at a shift of one σ.

5. CONCLUSIONS

Generalized control charting is an alternative to traditional $\bar{X}$ control charting. It provides sensitivity to various departures from the stationary, in-control process variable. It appears to be comparable to CUSUM. As with CUSUM, it avoids the "zone" approximations implicitly made in using $\bar{X}$ charts. Also, as with CUSUM, it avoids the use of ad hoc supplementary runs rules. As with CUSUM, it may be relatively hard to interpret, especially for the novice. Unlike CUSUM, it can be used for a process variable with arbitrary distribution—as long as its in-control distribution can be adequately characterized. Further analytical work and experience with GCC is needed to make a full comparison with existing methods.

The GCC procedure has potential use in other control chart situations. One obvious use is for the R-chart or s-chart. Since the distributions of R and s are known for arbitrary sample sizes, p-values could be computed for each subgroup variability statistic, and then the GCC method could be applied. The advantages would be the usual arbitrary interval length, no approximations of process variable level, and arbitrary α-level advantages enjoyed by ordinary GCC.

Another possible use is in the multivariate control setting. The complexity of multivariate control is such that it is common to use simple flagging rules for control statistics. For example, with a generalized Hotelling's T^2 chart, the group-mean statistic and group-variability statistic may be plotted separately, and commonly have a Shewhart-like "one-beyond-3σ" rule applied. Just as with the univariate $\bar{X}$ chart, R-chart or s-chart, p-values can be computed for the T^2 mean and dispersion statistics, and the GCC method can be applied.

APPENDIX A. CRITERIA FOR CONTROL CHARTS

The utility of this method can be demonstrated using carefully constructed synthetic data, where we can test the capability of a control chart procedure to meet certain criteria. Some criteria that seem sensible for a practical statistical control chart procedure are as follows.

1. The actual confidence level must be known. Preferably, it should be close to, if not equal to, the nominal level.
2. There must be a natural interpretation of "out-of-control" declarations. That is, the labels, "too low" and "too high" must make common English sense, and must mean that the sequence or individual value can be given some engineering interpretation.
3. The procedure must not be hard to implement on a computer.
4. The confidence level must be easily selected by the user.
5. The procedure must "immediately" identify extreme and sudden shifts in location.
6. The procedure must quickly identify moderate and slight, but persisting, shifts in location.
7. The procedure must quickly identify significant trends.
8. The procedure must have good power for commonly selected confidence levels, such as 90%, 95%, and 99%. [This is a more general restatement of (5)–(7).]
9. The procedure must work for virtually any parent distribution, assuming IID observations.
10. When observations are strongly serially correlated (a violation of the IID assumption), the procedure must flag out-of-control.
11. The procedure must work reasonably well in the case of slight serial correlation.

12. The procedure must not be unduly sensitive to small departures from the historical parent distribution—when the departures are not in the tails.

The ability of the GCC algorithm to satisfy criteria 1, 4, 5, 6, 7, and 8 was assessed empirically by the Monte Carlo runs described in Section 4. Analytic work is ongoing.

APPENDIX B. MOTIVATION FOR GENERALIZED CONTROL CHARTING

Traditional $\bar{X}$ control charting—as described by the AT&T *Statistical Quality Control Handbook* (1956 and later) and elsewhere—has some important positive and negative features. The positive features are well documented. Some of the negative features are:

1. The supplementary runs rules, though sensible, are ad hoc. Commonly, implementations incorporate runs rules ranging from one to eight units of time. Some people add a rule covering 20 units of time (17 out of 20 points on one side of the centerline). The individual confidence levels achieved by the rules are reasonable, and can be easily computed by anyone with a gaussian table and a hand calculator. In fact, anyone with this equipment can invent his own rules with any desired Type I confidence level, spanning any number of time intervals. The difficulty lies in selecting a set of rules, then computing the power of the set. One approach has been to "cover the spectrum" of anticipated time intervals by using a variety of rules—each oriented towards slightly different process behavior, but overlapping considerably. This is perhaps the best that can be done, but it is still ad hoc.
2. The rules are (conservative) approximations. For example, values just above the $\mu + \sigma$ line and just below the $\mu + 2\sigma$ line are treated identically, though their unlikelihood is different. Those adjacent to the lower line are at the 84.2% point, while those adjacent to the upper line are at the 97.7% point. The supplementary runs rules increase power for small shifts, but they still approximate with resolution no better than σ.

3. When you use sample means in an $\bar{X}$ chart, you intentionally lose the peculiar characteristics of the parent distribution from which you are drawing values. This is because the central limit theorem, in all but textbook situations, allows us to find a sample size large enough to get the distribution of sample means very close to the gaussian distribution—which is well understood. The disadvantage of plots of sample means is that the control chart user may become detached from the shape of the parent distribution. The parent distribution is a "signature" of the process, and ideally should be familiar to those who are trying to control the process.

APPENDIX C. TABLES OF ARLs FOR GCC AND CUSUM

This appendix contains tables of ARL values for three traditional CUSUM schemes and a large number of cases of GCC. CUSUM results were taken from published sources. The GCC results are all Monte Carlo estimates provided by the author, with standard error of the ARL shown. The tables show the ARLs for different values of p_{min} and k_{max}, under a variety of alternate hypotheses. Tables 1 and 2 show ARLs for p_{min} values of (10^{-2}, 10^{-3}, 10^{-4}, 10^{-5}, and 10^{-6}), shift and trend, respectively. Tables 3 and 4 show ARLs for p_{min} values of ($10^{-3.2}$, $10^{-3.4}$, $10^{-3.6}$, $10^{-3.8}$), also shift and trend, respectively. Each row of ARL values has a row under it containing the corresponding standard errors of the estimates. A plus sign indicates that the actual value is known to be higher, because a small proportion of the Monte Carlo runs did not terminate with an out-of-control flag. When this was the case for a moderate proportion of the runs, a double plus is used. When this was the case for a large proportion, the ARL was not reported. Note that in all cases of termination of at least half of the runs, the median run length *can be* obtained, unlike the ARL.

TABLE 1 ARLs for Shift, $p_{min} = (10^{-2}, 10^{-3}, 10^{-4}, 10^{-5}, 10^{-6})$

	Size of shift								
	0	0.25	0.50	0.75	1.00	1.50	2.00	2.50	3.00
h, k				CUSUM ARLs					
8, 0.25	730	85	29	16.4	11.4	7.1	5.2	4.2	3.5
−5, 0.5	930	140	38	17	10.5	5.8	4.1	3.2	2.6
2.5, 1	715	205	68	27	13.4	5.4	3.25	2.3	1.85
k_{max}, p_{min}				GCC ARLs/SE mean					
5 10^{-2}	77.2	40.7	16.7	10.57	7.43	3.23	2.02	1.660	1.320
	3.0	3.8	1.5	0.87	0.74	0.24	0.11	0.091	0.055
5 10^{-3}	615+	177+	67.9+	35.4	17.3	6.09	3.21	2.32	1.720
	24	15	5.4	3.5	1.5	0.49	0.17	0.10	0.074
5 10^{-4}	—	217+	93.5+	79.8+	42.6+	13.8	4.76	3.11	2.170
	—	26	8.4	6.2	3.6	1.2	0.30	0.14	0.091
5 10^{-5}	—	125+	82.9+	84.9+	76.2+	28.0	8.35	4.28	2.74
	—	41.3	17.9	9.9	6.0	2.3	0.71	0.22	0.13
5 10^{-6}	—	—	85.5	93.2+	100.2+	59.4+	13.7	6.21	3.56
	—	—	59.5	14.0	13.4	5.0	1.2	0.39	0.18
20 10^{-2}	71.9	36.0	13.3	9.44	6.16	3.13	2.02	1.660	1.320
	2.8	3.4	1.1	0.75	0.47	0.20	0.11	0.091	0.055
20 10^{-3}	538	133+	45.1+	19.3	12.01	5.15	3.13	2.32	1.720
	21	11	4.0	1.5	0.92	0.30	0.15	0.10	0.074

TABLE 1 (continued)

k_{max}	p_{min}	Size of shift: 0	0.25	0.50	0.75	1.00	1.50	2.00	2.50	3.00
					GCC ARLs/SE mean					
20	10^{-4}	2027++	206+	72.7+	32.2	17.0	7.55	4.25	3.08	2.170
		—	17	5.4	2.3	1.1	0.40	0.20	0.13	0.091
20	10^{-5}	—	201+	99.9+	57.6+	22.9	9.79	5.39	3.90	2.68
		—	37	8.4	4.1	1.4	0.47	0.23	0.14	0.11
20	10^{-6}	—	178+	97.5+	72.5+	36.9	12.28	6.62	4.88	3.36
		—	83	12.9	5.7	3.0	0.51	0.25	0.17	0.14
50	10^{-2}	71.6	35.7	12.89	9.42	6.16	3.13	2.02	1.660	1.320
		2.8	3.3	0.99	0.74	0.47	0.20	0.11	0.091	0.055
50	10^{-3}	529	120+	36.4	17.6	11.74	5.15	3.13	2.32	1.720
		20	10	2.8	1.2	0.83	0.30	0.15	0.10	0.074
50	10^{-4}	1982++	211+	61.9+	26.4	16.22	7.51	4.25	3.08	2.170
		—	15	4.3	1.5	0.95	0.38	0.20	0.13	0.091
50	10^{-5}	—	219+	83.1+	35.0	20.8	9.73	5.39	3.90	2.68
		—	26	5.6	1.8	1.1	0.45	0.23	0.14	0.11
50	10^{-6}	—	206+	85.5+	46.0	25.4	12.20	6.62	4.88	3.36
		—	50	6.9	2.3	1.2	0.48	0.25	0.17	0.14

TABLE 2 ARLs for Trend, $p_{\min} = (10^{-2}, 10^{-3}, 10^{-4}, 10^{-5}, 10^{-6})$

		Rate of trend								
		0	0.001	0.002	0.005	0.01	0.02	0.05	0.1	0.2
h,	k	CUSUM ARLs								
8,	0.25	730			77	52	35	21	15	10.6
5,	0.5	930	245	142	89	58	37	21	14.3	9.8
2.5,	1	715			96	64	41	23	14.3	9.5
$k_{\max}$,	$p_{\min}$	GCC ARLs/SE mean								
5	10^{-2}	77.2	54.8	59.8	45.8	36.7	24.4	15.95	10.87	6.89
		3.0	5.1	2.9	2.1	1.5	0.72	0.38	0.25	0.10
5	10^{-3}	615+	237	182.9	104.7	73.9	47.3	24.61	14.95	9.22
		24	15	9.3	4.4	2.8	1.5	0.77	0.39	0.24
5	10^{-4}	—	541	340	173.0	107.0	63.3	31.48	17.97	11.13
		—	13	6.1	2.8	1.5	0.75	0.41	0.24	0.10
5	10^{-5}	—	503	243.8	134.8	76.5	37.29	20.88	12.50	6.28
		—	10	6.0	3.1	1.5	0.67	0.36	0.21	0.11
5	10^{-6}	—	634	288.1	159.8	89.8	42.92	23.56	13.54	6.820
		—	13	6.0	2.7	1.5	0.70	0.38	0.22	0.096
20	10^{-2}	71.9	50.9	55.7	42.4	33.0	21.9	15.20	10.39	6.79
		2.8	4.6	2.7	1.8	1.3	0.66	0.35	0.24	0.10

TABLE 2 (continued)

k_{max}	p_{min}	Rate lf trend 0	0.001	0.002	0.005	0.01	0.02	0.05	0.1	0.2
		GCC ARLs/SE mean								
20	10^{-3}	538	203	150.2	86.7	60.3	38.0	21.13	13.88	8.79
		21	7.6	3.6	2.2	1.1	0.61	0.33	0.21	0.093
20	10^{-4}	2027++	387	238.6	128.7	79.2	47.3	25.49	15.75	10.25
		—	8.4	3.9	1.8	1.2	0.60	0.34	0.20	0.093
20	10^{-5}	—	344.1	161.1	93.4	54.2	28.21	17.66	11.31	6.160
		—	7.0	3.7	2.0	1.1	0.56	0.29	0.18	0.093
20	10^{-6}	—	398.7	187.4	106.6	62.0	30.72	19.21	12.34	6.740
		—	6.4	3.9	2.0	1.1	0.51	0.26	0.16	0.086
50	10^{-2}	71.6	50.9	54.0	42.4	33.0	21.8	15.20	10.39	6.79
		2.8	4.3	2.7	1.8	1.2	0.66	0.35	0.24	0.10
50	10^{-3}	529	196	142.7	79.0	57.2	36.3	20.98	13.88	8.79
		20	7.1	3.1	1.9	1.0	0.59	0.33	0.21	0.093
50	10^{-4}	1982++	325	210.1	109.2	73.7	44.2	25.13	15.75	10.25
		—	7.2	3.3	1.4	1.0	0.56	0.34	0.20	0.093
50	10^{-5}	—	274.7	135.7	82.8	49.56	27.93	17.66	11.00	6.160
		—	6.0	3.2	1.5	0.95	0.54	0.29	0.18	0.093
50	10^{-6}	—	318.1	153.4	90.2	54.73	30.28	19.21	12.34	6.740
		—	5.7	3.1	1.5	0.90	0.50	0.26	0.16	0.086

TABLE 3 ARLs for Shift, $p_{min} = (10^{-3.2}, 10^{-3.4}, 10^{-3.6}, 10^{-3.8})$

		Size of shift								
		0	0.25	0.50	0.75	1.00	1.50	2.00	2.50	3.00
h,	k				CUSUM ARLs					
8,	0.25	730	86	29	16.4	11.4	7.1	5.2	4.2	3.5
5,	0.5	930	140	38	17	10.5	5.8	4.1	3.2	2.6
2.5,	1	715	205	68	27	13.4	5.4	3.25	2.3	1.85
k_{max},	p_{min}				GCC ARLs/SE mean					
5	$10^{-3.2}$	918	177+	90.5+	45.6+	42.8+	7.73	3.71	2.34	1.820
		43	14	8.2	3.6	3.2	0.63	0.22	0.12	0.077
5	$10^{-3.4}$	1313+	194+	103.1+	52.3+	44.1+	8.51	3.81	2.50	1.930
		67	15	9.4	4.1	3.3	0.67	0.23	0.14	0.078
5	$10^{-3.6}$	1609+	184+	124+	56.8+	48.4+	9.65	4.23	2.63	2.060
		67	17	12	4.5	3.9	0.72	0.25	0.14	0.085
5	$10^{-3.8}$	1927+	196+	150+	60.5+	49.3+	10.79	4.64	2.86	2.120
		77	22	13	5.0	4.3	0.81	0.32	0.15	0.081
20	$10^{-3.2}$	802+	146+	51.8	20.9	18.8	5.71	3.56	2.32	1.820
		39	10	4.6	1.5	1.1	0.34	0.19	0.12	0.077
20	$10^{-3.4}$	1108+	166+	63.1	25.8	20.2	6.09	3.64	2.44	1.930
		50	12	5.9	1.9	1.2	0.34	0.19	0.12	0.078

TABLE 3 (continued)

		Size of shift								
		0	0.25	0.50	0.75	1.00	1.50	2.00	2.50	3.00
k_{max}, p_{min}					GCC ARLs/SE mean					
20	$10^{-3.6}$	1393+	185+	74.8	29.6	21.6	6.53	3.94	2.58	2.060
		60	14	7.2	2.0	1.2	0.35	0.20	0.13	0.085
20	$10^{-3.8}$	1762+	190+	83.2	35.3	24.3	7.11	4.15	2.78	2.120
		71	16	7.5	2.7	1.6	0.36	0.21	0.13	0.081
50	$10^{-3.2}$	792	131.4	42.8	19.2	17.13	5.71	3.56	2.32	1.820
		39	9.2	3.2	1.2	0.74	0.34	0.19	0.12	0.077
50	$10^{-3.4}$	1077+	153+	48.2	21.9	18.11	6.09	3.64	2.44	1.930
		50	10	3.7	1.3	0.77	0.34	0.19	0.12	0.078
50	$10^{-3.6}$	1335+	171+	53.0	24.3	19.17	6.53	3.94	2.58	2.060
		59	11	4.0	1.3	0.80	0.35	0.20	0.13	0.085
50	$10^{-3.8}$	1680+	186+	56.0	26.0	20.24	7.11	4.15	2.78	2.120
		67	13	4.2	1.4	0.89	0.36	0.21	0.13	0.081

TABLE 4 ARLs for Trend, $p_{min} = (10^{-3.2}, 10^{-3.4}, 10^{-3.6}, 10^{-3.8})$

		Rate of trend								
		0	0.001	0.002	0.005	0.01	0.02	0.05	0.1	0.2
		CUSUM ARLs								
h,	k									
8,	0.25	730			77	52	35	21	15	10.6
5,	0.5	930	245	142	89	55	37	21	14.3	9.8
2.5,	1	715			96	64	41	23	14.6	8.5
		GCC ARLs/SE mean								
k_{max},	p_{min}									
5	$10^{-3.2}$	918+	309	214.7	129.0	78.4	48.8	25.46	15.31	9.52
		43	17	9.6	5.0	3.3	1.7	0.74	0.38	0.28
5	$10^{-3.4}$	1313+	371	247	141.2	85.8	52.1	26.62	16.16	10.07
		67	19	10	4.9	3.3	1.7	0.68	0.39	0.27
5	$10^{-3.6}$	1609+	427	278	154.5	92.1	54.6	28.04	16.92	10.37
		67	21	11	5.4	3.2	1.8	0.63	0.40	0.27
5	$10^{-3.8}$	1927+	506	316	169.0	100.3	58.0	29.95	17.68	10.63
		77	22	12	5.4	3.1	1.8	0.63	0.40	0.25
20	$10^{-3.2}$	802+	227	169.6	103.1	61.7	38.5	21.57	13.97	8.67
		39	12	7.3	4.1	2.5	1.1	0.62	0.34	0.24
20	$10^{-3.4}$	1108+	262	184.4	111.8	69.1	40.6	22.67	14.53	9.12
		50	13	7.5	3.8	2.7	1.1	0.56	0.31	0.22

TABLE 4 (continued)

		Rate of trend								
		0	0.001	0.002	0.005	0.01	0.02	0.05	0.1	0.2
k_{max}, p_{min}		GCC ARLs/SE mean								
20	$10^{-3.6}$	1393+	305	204.2	118.5	71.7	42.5	32.65	14.88	9.50
		60	14	7.4	3.8	2.6	1.2	0.53	0.31	0.24
20	$10^{-3.8}$	1762+	346	227.5	125.7	78.4	44.4	24.40	15.45	9.81
		71	16	7.7	3.4	2.4	1.2	0.53	0.31	0.22
50	$10^{-3.2}$	792	217	155.7	92.5	58.9	37.1	21.55	13.97	8.67
		39	12	6.4	3.3	2.3	1.1	0.62	0.34	0.24
50	$10^{-3.4}$	1077+	243	166.3	101.5	63.1	39.5	22.64	14.53	9.12
		50	12	6.5	3.3	2.4	1.1	0.56	0.31	0.22
50	$10^{-3.6}$	1335+	273	179.7	107.7	65.2	40.9	23.48	14.88	9.50
		59	12	6.5	3.3	2.3	1.0	0.51	0.31	0.24
50	$10^{-3.8}$	1680+	303	199.2	113.0	70.5	42.3	24.25	15.45	9.81
		67	13	6.3	3.0	2.1	1.0	0.51	0.31	0.22

REFERENCES

Abramowitz, M., and I. Stegun. (1972). *Handbook of Mathematical Functions with Formulas, Graphs, and Mathematical Tables*, pp. 931, 958–959. Wiley, New York.

Asbagh, N. A. (1985). Performance of CUSUM and combined Shewhart–CUSUM charts under linear trend. Unpublished paper at the University of Southwestern Louisiana.

AT&T. (1956 and later). *Statistical Quality Control Handbook*, pp. 25–27. AT&T, Indianapolis, Indiana.

Bissell, A. F. (1984). The performance of control charts and Cusums under linear trend. *Applied Statistics*, *33*(2), 145–151. [See corrections in Bissell (1985).]

Bissell, A. F. (1985). Letter to the editor, *Applied Statistics*, June 10, 1985.

Goel, A. L., and S. M. Wu. (1971). Determination of ARL and a contour nomogram for Cu-Sum charts to control normal mean. *Technometrics*, *13*(2), 221–230.

Johnson, N. L. and S. Kotz. (1970). *Distributions in Statistics: Continuous Univeriate Distributions*, Vols. 1 and 2, Wiley, New York.

Lucas, J. M. (1973). A modified V mask control scheme. *Technometrics*, *15*(4), 833–847.

Lucas, J. M. (1982). Combined Shewhart–CUSUM quality control schemes. *Journal of Quality Technology*, *8*(1), 1–12.

Lucas, J. M., and R. B. Crosier. (1982). Fast initial response for CUSUM quality control schemes: Give your CUSUM a head start. *Technometrics*, *24*(3), 199–205.

Quality Magazine. (1986). Software update. December, pp. 50–63.

Wadsworth, H. M., K. S. Stephens, and A. B. Godfrey. (1986). *Modern methods for quality control improvement*, pp. 233–248, Wiley, New York.

Woodall, W. H. (1983). The distribution of the run length of one-sided CUSUM procedures for continuous random variables. *Technometrics*, *25*(3), 295–301.

SECTION IV

Statistical Databases for Process Control

10
Statistical Databases for Automated Manufacturing

Sakti P. Ghosh
IBM Almaden Research Center
San Jose, California

Statistical process control has taken on a new dimension with the introduction of microprocessors in instrumentation and processing of manufacturing data. Classical statistical quality control based on batch processing is being replaced by automated statistical process control. This paper addresses origin, utilization, storage, and benefits of statistical databases in automated manufacturing systems. A data model containing category attributes and statistical attributes is presented for statistical databases. Some specific features of automated manufacturing processes are characterized in terms of data represented by this model for statistical databases. Differences between this model and the classical relational database model are discussed. Data retrieval schemes from the proposed models using SQL/DB2 are also discussed. Some data structures

Based on a presentation made at "Statistical Process Control: Keeping Pace with Automated Manufacturing, a National Symposium," sponsored by the Center for Professional Development and the Reliability, Availability and Serviceability Laboratory, College of Engineering and Applied Sciences, Arizona State University, November 6–7, 1986.

based on a shadow data set, suitable for manufacturing statistics, are also covered. Similarities between such data structures and statistical martingale sequences are discussed. Some discussion on the role of manufacturing information on cost is also provided.

1. INTRODUCTION

In modern manufacturing processes, computers are used extensively in almost every phase. Computers generate and process various types of information. In automated manufacturing, the computers control the different manufacturing subprocesses based on programmed logic with minimum human intervention. The intelligence needed to update the process control rules, parameters, and decisions are based on the data generated by the manufacturing system. The data set generated by the manufacturing system is called manufacturing information. The software system that processes such information is called a *manufacturing information system* (MIS). Some of this information is category type of information, such as parts number or process step number. Sometimes the information is statistical in nature, such as values of test parameters, or measurements. Sometimes the information is pictorial in nature, such as diagrams of parts design, or manufacturing shop floor plans. Most of the available *database management systems* (DBMS) are very efficient in processing category information. Most of the research in DBMS in the last 20 years has been focused in this effort. Almost all DBMSs can process statistical information, but they do it very inefficiently. The reason for this is that statistical processing of information involves processing a large volume of data, which takes time. Most statistical queries involve category-statistical processing, for example, what is the average diameter of the bolts manufactured by Workstation 10?

Recently, there has been considerable interest in efficient processing of statistical information. This phenomenon has been accelerated by the fact that the cost of manufacturing labor is increasing very rapidly. To hold down cost, computerized manufacturing is being introduced at a very fast pace. Many researchers in statistical database management have been examining these problems (Ghosh, 1983, 1985b; Hammond and McCarthy, 1983; Wong, 1982a). Though researchers have made some progress in the understanding of statistical databases, the use of statistical databases in computer-generated manufacturing information systems has not advanced much. Usually the so-called statistical data are actually a combination of category and statistical information. Most of the present research in statistical information has concentrated on the category information aspect of statistical databases. Some of the good work that can be cited in statistical databases is in the area of modeling (Shoshani, 1982; Su, 1983; Johnson, 1981; Kreps, 1982), which reflects the work done at University of California, Berkeley (see Wong, 1982a) and the University of Florida at Gainsville (see Su, 1983). Three good workshops on statistical databases have been organized, and the proceedings of three workshops sum up the best work done in this field up to 1986 (Wong, 1982b; Hammond and McCarthy, 1983; Cubitt et al., 1986). Most of the work in statistical databases for manufacturing have followed since then (Ghosh, 1983, 1985a, b, c; Su, 1986).

Availability of instant statistical information can play a major role not only in continual flow manufacturing process control but also in the dynamic controlling of test execution based on the actual result of the manufacturing process. This can lead to major reduction of the cost of computerized testing (Ghosh, 1985b), inventory control, statistical process control, etc. It is true that many statistical process control methods use operation research techniques extensively, but most of the techniques are not based on processing of statistical information in real time. In some manufacturing processes (e.g., chemical depositions) instant processing of statistical information may not be a critical factor, but in computerized testing or continuous flow manufacturing, real-time statistical estimation of parameters is critical. Present data access methods (ISAM, VSAM, XRAM, etc.) are efficient in accessing category information, but are not efficient in accessing statistical information. Ghosh (1985c) has attempted to attack this problem, and his work may motivate other researchers to find better solutions to this problem.

Most manufacturing processes are concerned with producing some objects that have a solid form; information production is not the main goal of most of the manufacturing work stations. There are two basic types of manufacturing information: (1) process planning data and (2) business planning data. Process planning data (also referred to as technical data, operation process data, process control data, etc.) originate by testing products, monitoring of equipments, from analytic models, simulation, etc. Statistical quality control data, data for *computer-aided design* (CAD), and data for *computer-aided manufacturing* (CAM) all belong to this category. Business planning data relate to the mostly business side (economics) of manufacturing. Data related to bill of materials, cost of production, volume of production, due dates, inventory, etc. belong to this category of manufacturing information. Unlike process planning data, some business planning data can originate outside the manufacturing environment, for example, economic surveys and reports. Most of these data are category-statistics data, but sometimes they may also include pictorial data (data from a scanner are also included in this type), such as statistical charts, quality control charts, diagrams of product design, or floor plan of an automation process.

One of the fundamental goals of all manufacturing information is how to use it to reduce cost of manufacturing. This may be achieved through such means as automation of the manufacturing process, reduction in idle time of manufacturing devices, production of better quality products and thus reduction of cost of rejected units, reduction of inventor tax, or competitive pricing of products. In all these approaches, timely availability of the correct type of information can benefit the manufacturing process to a great extent. In this paper we will be concerned with this problem. In particular we will focus on processing, storing, and retrieval of the statistical component of category-statistics manufacturing information. In the next section some statistical database structures needed for automated manufacturing are discussed. In Section 3 manufacturing statistics, which can have a major impact on the cost of manufacturing, are covered. In Section 4, manufacturing information economics are discussed. Some concluding remarks are covered in Section 5.

2. STATISTICAL DATABASE STRUCTURES

Repetitiveness is inherent in manufacturing (the same component type is manufactured over and over again), and statistics is the science of averages: thus there is a natural binding between these two areas. Though often the characteristics of an individual component need to be examined carefully, the ultimate goal of every manufacturing process is to achieve a high level of average excellence. Thus statistics plays a very important role in manufacturing. The use of statistics in manufacturing is not new: statistical quality control methods and sampling techniques for manufacturing have been in use since the early 1940s (see Bell Labs, 1982). The new dimensions that microprocessors have introduced are the creation of a very high volume of statistical data and the need for instant statistical analysis and feedback to the automated manufacturing system. Most of the present so called *statistical databases* contain records of individual units. Any statistical analysis based on such statistical databases is extremely slow; thus researchers in this area have suggested the storage of statistical aggregates in the databases (Ghosh, 1983; Shoshani, 1982). Another basic structure in manufacturing databases (and in most practical databases) is that they contain not only statistical information but also category and pictorial information. In this paper, properties and structures of pictorial information are not discussed. Recently some good work on database design for manufacturing systems has been done at the University of Florida (Navathe et al., 1986).

One of the basic goals of any manufacturing database is to collect information that is essential for performing the manufacturing process, improving the quality of the items manufactured, reducing the cost of manufacturing, and providing feedback for better product design. If all these are to be achieved, then the manufacturing database would wind up having all the necessary information for simulating (completely) all the phases of the manufacturing process; however, the cost of such an information system would be prohibitive and its true usefulness to the manufacturing system is questionable. Thus the problems are what manufacturing information should be collected and how the manufacturing information can be best utilized. These are very difficult questions; operation research scientists and industrial engineers have been looking into these problems for many years. Statisticians have also joined them in this effort. Now the computer scientists have joined the team also. Some of the major sources of statistical information in a manufacturing system are the testing and inspection of products. Statistical information are also generated from process control measurements, business planning, etc. Automated manufacturing systems are evolutionary in nature, i.e., they are changing gradually but continually, and so also is the information generated by them. Hence any good manufacturing statistical database has to be highly adaptive to changes. Codd's (1970) relational model for databases is ideal for a manufacturing environment due to the large number of sources at which manufacturing information may originate and also the continual change in the logical associations between these databases. The relational model along with the commercial products based on a relational model (SQL/Data System Application Programming, 1983) can be a very important resource for constructing databases for manufacturing information systems associated with automated manufacturing.

The category attributes of the databases associated with any manufacturing system relate to the different resources involved in the manufacturing operations. Thus the category attributes can relate to names of the components manufactured, the item ids (identifications), the names of the manufacturing operations (e.g., grinding, electrolysis, Station X), the manufacturing device ids that execute the manufacturing operations on the units, date–time when the manufacturing operations were executed, the human operators (if any) who were involved in the execution of the manufacturing operations, etc. Let us denote the category attributes as $A_1, A_2, \ldots, A_n$.

The statistical attributes of the manufacturing databases are those attributes that relate to repetitive numerical measurements performed on the resources at any state of the manufacturing operations. Examples of such attributes are parametric measurements on manufactured components, numeric outcome of manufacturing tests, cost information (e.g., price of resources, cost of processing information) associated with different resources at different states of manufacturing operations, etc. Let us denote the statistical attributes as $S_1, S_2, \ldots, S_m$.

One of the basic differences in usage of the information contained in the category attributes $A_1, A_2, \ldots, A_n$ and the statistical attributes $S_1, S_2, \ldots, S_m$ is that, when tuples of the category attributes contain repeating values of attribute subsets it usually reflects an embedded hierarchical structure [also known as repeating groups in relational database; see Roth et al. (1985), Ghosh

(1986)], where as when repeating values of attribute subsets appear in the statistical attributes they are replaced by one tuple and a count of the number of occurrences of the tuples [this generates the multivariate frequency distributions; see Kendall and Stuart (1958)]. These two approaches of data reduction for the two types of attributes are basic steps, which lead to usage of different types of algebra and calculus. In the case of category attributes, set theory, relational algebra, relational calculus, and network theory are applicable for information manipulation. In the case of statistical attributes, number theory, numerical calculus, differential calculus, and multidimensional geometry are used for processing statistical information. In statistical databases, numerical operations can be performed between the values of the attributes and their frequencies of occurrences (e.g., calculations of means, standard deviations, etc.), which are not possible with category attributes. It would be nice if an algebra or a calculus could be developed that could be used for both category and statistical attributes for extracting logical structural and statistical information efficiently. This topic is outside the scope of the present paper and is left for future researchers.

Let us denote the cost function of the events that specify the values of these attributes as $c(A_1, A_2, \ldots, A_n, S_1, S_2, \ldots, S_m)$. In order to specify a realization of this cost function, we have to assign a specific value to each of the attributes in the function. We will now decompose stepwise this cost function into major components. In the first step the cost function is decomposed into (1) cost related to price of the resource and (2) the cost for utilizing the manufacturing information associated with the resource. Thus we have

$$c(A_1, A_2, \ldots, A_n, S_1, S_2, \ldots, S_m)$$
$$= c(A_1, A_2, \ldots, A_n) + c(U(A_1, A_2, \ldots, A_n, S_1, S_2, \ldots, S_m)) \quad (1)$$

where $c(A_1, A_2, \ldots, A_n)$ is the price associated with the resources specified by the event, and $c(U(A_1, A_2, \ldots, A_n, S_1, S_2, \ldots, S_m))$ denotes the cost associated with the utilization of the statistical information pertinent to the resources, for the manufacturing event. The cost component associated with the utilization of the statistical information can be further decomposed into the cost associated with (1) the retrieval of the information from the computer storage and (2) the processing of the information by the central processing unit (CPU), which is given next:

$$\begin{aligned} c(U(A_1, A_2, \ldots, A_n, S_1, S_2, \ldots, S_m)) \\ = c(\mathrm{RI}(A_1, A_2, \ldots, A_n, S_1, S_2, \ldots, S_m)) \\ + c(\mathrm{CPU}(A_1, A_2, \ldots, A_n, S_1, S_2, \ldots, S_m)) \end{aligned} \tag{2}$$

where $c(\mathrm{RI}(A_1, A_2, \ldots, A_n, S_1, S_2, \ldots, S_m))$ is the cost for retrieving the relevant information for the manufacturing event, and $c(\mathrm{CPU}(A_1, A_2, \ldots, A_n, S_1, S_2, \ldots, S_m))$ is the cost for the CPU to process the retrieved information. In utilizing manufacturing information, that is, $U(A_1, A_2, \ldots, A_n, S_1, S_2, \ldots, S_m)$, sometimes some of the numeric information are utilized in the form in which they were observed by the manufacturing system, and sometimes the statistical contents in a large volume of data are used.

Both $c(\mathrm{RI}(A_1, A_2, \ldots, A_n, S_1, S_2, \ldots, S_m))$ and $c(\mathrm{CPU}(A_1, A_2, \ldots, A_n, S_1, S_2, \ldots, S_m))$ are monotonic increasing function of the volume of data; volume of data is directly related to access time, processing time, and cost. Thus, the structure of the data contained in $\mathrm{RI}(A_1, A_2, \ldots, A_n, S_1, S_2, \ldots, S_m)$ should be of the form of *statistics metadata* (see Ghosh, 1985a), which also minimizes the cost component $c(\mathrm{CPU}(A_1, A_2, \ldots, A_n, S_1, S_2, \ldots, S_m))$ subject to other constraints. The structures of the statistics metadata discussed in the cited paper do minimize the retrieval time, hence that topic is not covered again. Some of the other DBMS constraints, such as updating of data, disaggregation of statistical data for subcategory structures, and integrity constraints, have also been discussed in the cited paper. These statistics metadata also reduce the cost of data communication in a computer network environment.

Manufacturing data are subjected to two major types of usage: (1) routine manufacturing process control, and (2) nonplanned usage, e.g., defect diagnosis, new product development, etc. Thus for designing generic databases for manufacturing information systems, it is prudent to choose the database design such that the performance is reasonably good for a large class of applications. Hence we advocate the relational model. DBMS based on the relational model (e.g., SQL/DB2) provide reasonably good performance for dynamic logical structures created by end-users on category attributes. The same performance claims usually are not valid for statistical databases with category and statistical attributes. However, if the values of the statistical attributes are stored as statistical aggregates (as in statistics metadata), the performance improves considerably. If all the queries are based on the category attributes only, then the choice of the granularity of the statistical summarization is simple. In such a situation, the granularities are the category attribute atomic cells. Within each cell, different statistical aggregates (sums, sums of squares, sums of products, multivariate frequency distributions, etc.) can be stored. If the qualification part of the query structures also contains statistical attributes, then granularities of the statistical summarizations are dependent on the queries themselves. Thus a typical relational structure for the statistical databases may be of the form

$$R(A_1, A_2, \ldots, A_n, \mathrm{AG}(S_1), \mathrm{AG}(S_2), \ldots, \mathrm{AG}(S_m)) \qquad (3)$$

where the $\mathrm{AG}(S_i)$ terms are the statistical attributes with complex relational domain for the statistical aggregates. In many queries the qualification part is based on a set of values of category attributes, e.g., "Give me the frequency distribution of the lengths of part number 1325, manufactured by company X over the last five years." For such queries the additivity of the values of the attributes $\mathrm{AG}(S_i)$ over the category attribute cells is not automatically guaranteed unless the statistical aggregates themselves satisfy the property of additivity. Thus the construction of the statistical aggregates plays an important role in the performance of the queries directed at such statistical relational data structures.

In manufacturing processes where discarding of bad components is too expensive (also for defect diagnosis), numerical data relevant to individual units need to be preserved for some time. Some research in understanding statistical summarization of manufacturing data has been published recently (Ghosh, 1985b), and the subject is not repeated here.

The category attributes of the manufacturing system can be classified into four major groups: (1) product-related attributes, (2) chronology-related attributes, (3) equipment-related attributes, and (4) human resource-related attributes. The product-related attributes refer to characteristics of the units that are the output of the manufacturing equipment. The chronological attributes are time-related, e.g., date, clock, etc. The equipment attributes refer to characteristics of the hardware equipment that perform some role in the manufacturing of the product or its components. Attributes of monitoring instruments or test equipments also belong to this group. Computers attributes also belong to this group. Human resource attributes characterize the human effort involved in the manufacturing processes. These attributes could be as simple as the names of the manufacturing operators, or as complex as the task performed by a human operator on a product or cost of administrative overheads.

Example 2.1. Let us consider a statistical database for manufacturing information systems and discuss some typical examples of usage of such databases. It is assumed that the database is stored as a relational structure in the form given in Eq. (4), and is organized by a SQL/DB2 DBMS

$$R(A_1, A_2, \ldots, A_n, S_1, S_2, \ldots, S_m) \tag{4}$$

Sometimes data from different manufacturing lines (a manufacturing line consists of multiple workstations performing the same tasks on the same types of units) may be stored in different tables (relations). Thus join operations are needed for retrieval of data from multiple relations (i.e., tables). As the join operation is a well-known operation and does not affect the results that are discussed in this example, it is assumed that the necessary join operations have already been performed (to create views) and the relevant information is contained in the relation (table) given in Eq. (4).

Component traceability. In a manufacturing assembly line when a component is detected to be defective, the manufacturing engineers are usually interested in examining the data that were collected on the particular component at previous workstations. The relevant data can be retrieved by executing the following SQL/DB2 statement:

```
SELECT (*)
  INTO R2
    FROM R
      WHERE COMPONENT-ID = 'xxxxx';
```

The table R2 will contain all the data pertinent to the desired component "xxxxx." Various types of statistical and categorical analysis can be performed on the data contained in R2 to reveal many engineering and process control events of the manufacturing system, such as, (1) what are the values of the equipment attributes when the component xxxxx was manufactured? (2) what are the names of the operators who worked on the component xxxxx? (3) what was the average time for manufacturing the component xxxxx? (4) what is the standard deviation of a particular statistical attribute that was measured repetitively on the component xxxxx? etc.

Equipment monitoring. Most of the manufacturing equipment is electro-mechanical devices, and under heavy usage these do wear out and slowly drift away from their desired performances. Thus, most of the manufacturing equipment needs to be monitored and calibrated often. The monitoring of equipment is performed by statistically analyzing [see analysis of variance for testing equality of means in Box et al. (1978)] the data generated by the various testers. In order to retrieve the relevant data in the appropriate structural form from the table [Eq. (4)], the following SQL/DB2 statement needs to be executed:

```
DO WHILE (loop for all EQUIPMENT-IDs);
  SELECT (*)
    INTO R2(zzzzz)
      FROM R
        WHERE EQUIPMENT-ID = 'zzzzz'
          AND TIME between t1 and t2;
          NEXT zzzzz;
          END OF DO-loop;
```

[Please note that R2(zzzzz) represents a table name and not a tuple of the relation R2.] It is possible to include some statistical aggregations in the above SQL/DB2 statement, but we are not discussing statistical analysis to that level of detail.

Productivity of operators. In many manufacturing processes, the management is very interested in monitoring the productivity of the manufacturing operators. This information is contained in the statistical data for the number and quality of units manufactured by the different operators in different time intervals. Such information may be obtained from the table [Eq. (4)] by executing the following SQL/DB2 statement:

```
DO WHILE (loop for all OPERATOR-IDs);
  SELECT (*)
    INTO R2(yyyyy)
      FROM R
        WHERE OPERATOR-ID = 'yyyyy'
          AND TIME between t1 and t2;
          NEXT yyyyy;
          END OF DO-loop;
```

The data extracted by the above SQL/DB2 statement is too gross. Sometimes a count of the number of components from each of the tables R2(yyyyy) may be adequate for computing the productivity factor of the operator. Sometimes more detail may be warranted.

Many researchers have discussed the problem of repeating tuples in the context of relational databases for category attributes. These repeating tuples can contain embedded hierarchical structures [see Roth et al. (1985); G-relations in Su (1983)]. Most of these logical structures can be derived from the structural patterns in the values of the category attributes. In statistical databases the same structural pattern can repeat multiple times and thus the relational data structure given in Eq. (3) can be augmented to include a count attribute, whose value reflects the frequencies of occurrences of the particular pattern (of the values of the category attributes $A_1, A_2, \ldots, A_n$). This relational data structure is given in Eq. (5).

$$R(A_1, A_2, \ldots, A_n, \text{COUNT}, \text{AG}(S_1), \text{AG}(S_2), \ldots, \text{AG}(S_m)) \quad (5)$$

The attribute COUNT is a category-derived numerical attribute. When statistical data is represented in the form given in Eq.

(5), it is possible to compute statistical measures of associations between the values of the category attributes (see Kendall and Stuart, 1961, Chapter 33). The topic of statistical measures of associations has been developed extensively by statisticians, and is not repeated here. Readers interested in the subject may refer to the cited references. A useful example of such statistical category association in a manufacturing environment is discussed next.

Example 2.2. Suppose a manufacturing company is interested in studying if there is any statistical association between the different manufacturing shifts and the workstations. Such measures of associations between the category attributes can be computed from the tuples of the relation of Eq. (5). In order to compute such statistical measures of associations, a contingency table for the frequencies of occurrences of the different categories has to be extracted from the table given in Eq. (5). The SQL/DB2 statement that can achieve this is given next.

```
DO WHILE (loop for all MANUFACTURING-SHIFTs);
  DO WHILE (loop for all WORKSTATIONs);
    SELECT COUNT
      INTO R3.A(xxxxx, yyyyy)
        FROM R
          WHERE MANUFACTURING-SHIFT = 'xxxxx'
            AND WORKSTATION = 'yyyyy';
            NEXT xxxxx;
      END OF DO-loop;
    INITIALIZE xxxxx;
    NEXT yyyyy;
    END OF DO-loop;
```

(Please note that here R3.A represents a two-dimensional array.) It should be noted that SQL/DB2 has a build-in count function, but having a count attribute in the table improves performance considerably. The data that are extracted by the SQL/DB2 statement are then used to compute the value of the statistical measure of association. If necessary, the usual statistical test of hypothesis can also be performed (Rao, 1952; Bell Labs, 1982).

3. MANUFACTURING STATISTICS AND DATA STRUCTURE

In this section we shall discuss some of the structures of statistics, which can be important in automated manufacturing environment. In most manufacturing steps some of the basic techniques of statistics (e.g., estimation, testing of hypothesis) can be utilized effectively. Specifically, the following areas of manufacturing can benefit from statistical methods: statistical process control parameters (includes statistical quality control, floor shop control), product testing (includes defect diagnosis, dynamic sampling, test results prediction in real time, instrument calibration, instrument homogeneity testing, etc.), inventory/production control (includes business volume prediction, manufacturing volume prediction, cost prediction, etc.), product design (includes structural and behavioral prediction of parameters of new products, market volume prediction, cost prediction, etc.), .

In a manufacturing environment (as in most businesses), time is money. Thus the ability to compute the statistics very fast is a desirable feature for the usefulness of any statistics in a manufacturing line. Some statistical databases containing statistical aggregates can provide such properties (Ghosh, 1983, 1984). Ability to make instant decisions bases on real-time statistical computations is also important for high product quality and low cost. Like many practical systems the manufacturing systems are not perfect; thus they often produce bad data. It is very difficult to identify bad data in real time; thus, these data tend to be included in the manufacturing statistical computations. Hence another desirable property of a manufacturing statistics is its robustness (Kendall and Stuart, 1961) under bad data and minor drifting (changes) of statistical distributions. Many manufacturing processes are not true stationary processes; even when the manufacturing line is under statistical quality control they tend to change slowly. Thus, computational techniques based on the concept of *sliding windows* are more effective. Real-time statistical computations for such techniques need a statistical database containing special types of statistics metadata. The statistical database structure provided in the cited paper (Ghosh, 1983) has this property. It is possible that many other data structures may also have this flexibility, but the subject needs to be investigated. In a manufacturing environment, data are continually generated. Thus another desirable property that manufacturing statistics should have is the asymptotic property of convergence, i.e., as the number of observations increases, the statistics should become a better estimator of the parameter that it is supposed to estimate.

There are also some properties that are undesirable for manufacturing statistics. If the computation of the statistics is based on permutation or sorting or ordering of the observations in any manner, then the computation of the statistics is very time-consuming. Thus they are not appropriate for real-time manufacturing decisions.

Suppose t_n is a statistical estimate of the manufacturing parameter θ based on n observations. The asymptotic property of statistical convergence of t_n to θ can be stated as

$$\operatorname*{Lt}_{n \to \infty} \Pr\{t_n \to \theta\} \sim 1 \tag{6}$$

In order to achieve this in a manufacturing process, stationarity of the process (over time) is important. This cannot be guaranteed even for a manufacturing line under statistical quality control. Often stationarity over a short interval of time is attainable. Thus probabilistic asymptotic convergence can be achieved by sliding the window of reference. Many statistics satisfy the probabilistic asymptotic convergence property given in Eq. (6). Some of them are mean, statistical moments, relative frequency distributions, and product moments. However, these statistics do not satisfy the additive property under additional observations, which is given by

$$t_{n+1} = t_n + f(x_{n+1}) \tag{7}$$

where $f(x_{n+1})$ is some function of the $(n+1)$th observation x_{n+1}. Many statistics with probabilistic asymptotic convergence property do not have the additive property for additional observations but the following additive formula does hold:

$$t_{n+1} = t_n \frac{n}{n+1} + \frac{f(x_{n+1})}{(n+1)} \tag{8}$$

Analytically, Eq. (8) does provide the necessary additive property, but in practical computation, it can accumulate rounding errors causing a bias in the estimates.

In computerized manufacturing large amounts of data are generated continually and the life utility of most of the raw data diminishes very rapidly. Also the manufacturing process changes slowly and continually (slow-moving stochastic process). Hence the domain of utility of the data is a small window. In most situations this *utility window* is a sliding snapshot of the data over a time (most recent history). However, in many situations the window may define category objects such as "tools installed in a manufacturing plant," "units manufactured by certain groups of operators," etc. We have discussed extensively the importance of statistical aggregates in statistical databases. It is difficult to include the sliding window updating in such databases, because when data is aggregated it cannot be disaggregated again. To achieve this important property of utility window the data sets within the database have to contain some redundant data. The proposed database will contain two data sets: (1) a shadow data set and (2) the statistical data set containing statistical aggregates or metadata with an additive property.

Shadow data set. This data set may contain raw data or cleaned raw data (i.e., raw data after eliminating bad data, correcting for missing data, altering the data structure or format, etc.), or data summarized into some atomic structure (e.g., readings from multiple instruments on the same unit). The shadow data set is ordered with respect to the values of an ordering variable. The ordering variable may be the "time of origin of the data," "a generated sequencing attribute" (e.g., date of manufacturing, classification codes, product style, etc.), or any other ordering variable. The ordering variable in a shadow data set is used to define the boundaries and the sliding rules of the window or the foot-print of the statistical database. Let $R(A_1\ A_2, \ldots, A_n, S_1, S_2, \ldots, S_m)$ be the structure of the shadow data set. We have assumed that the structure of the shadow data and that of the raw data are the same. If this is not true then the attributes will represent the atomic values of the shadow data set. Usually one of the category attribute will represent the ordering variable.

Suppose $X_1, X_2, \ldots$ are ordered random variables. If

$$X_n \overset{\text{a.s.}}{=} E(X_{n+1} \mid X_1, X_2, \ldots, X_n)$$

[i.e., the expected value of the random variable X_{n+1} given the past history $X_1, X_2, \ldots, X_n$ is a.s. (almost surely) equal to the value of the last random variable X_n] then the sequence of the random variables $X_1, X_2, \ldots$ form a *martingale sequence* (see Loeve, 1960). Mathematicians and probabilists have studied martingales for some time, and the behavior of the manufacturing statistical databases (over time or manufacturing batches, etc.) has considerable similarities with them. Some of these properties are discussed here. An important martingale structure that is relevant for manufacturing processes is stated next:

If (1) $Y_i (i = 1, 2, \ldots)$ is a set of random variables, which are probabilistic independent, and $E(Y_i) = 0$, or (2) $Y_i (i = 1, 2, \ldots)$ is probabilistic dependent and $E(Y_k \mid Y_1, Y_2, \ldots, Y_{k-1}) \overset{\text{a.s.}}{=} 0$ then the random variables

$$X_i = \sum_{j=1}^{i} Y_j$$

form a martingale sequence.

Let us denote the random variable Y_i to represent the value (tuple) of the relation $R(A_1, A_2, \ldots, A_n, S_1, S_2, \ldots, S_m)$ as observed from a manufacturing process. We assume that each of the statistical variables $S_1, S_2, \ldots, S_m$ are measured from their mean, and thus the conditions $E(S_i) = 0$ are satisfied. We shall not impose any probabilistic constraint on the attribute variables $A_1, A_2, \ldots, A_n$. The only assumption made is that the union of the values of these attributes is finite. Thus the condition that the Y_i variables are probabilistic independent with $E(Y_i) = 0$ (as extended to vectors) is applicable when the manufacturing process is in control. The condition that the Y_i variables are probabilistic dependent with $E(Y_k \mid Y_1, Y_2, \ldots, Y_{k-1}) \overset{\text{a.s.}}{=} 0$ is applicable when the manufacturing process is not stable. In either case the aggregates generated from the observed tuples (by aggregating each attribute and variable separately) of the relation $R(A_1, A_2, \ldots, A_n, S_1, S_2, \ldots, S_m)$ form a martingale sequence. It is easy to develop the rules for determining which data elements should be included in the shadow data set based on the *σ-field* generated by the martingales (see Loeve, 1960). The σ-fields generated by the martingales are also the key to the determining which attributes should be included in the different relational databases (discussed in the previous section), so that properties of component traceability in the manufacturing process are preserved. The σ-fields generated by the martingales associated with the attributes in the different relational databases can also be useful in determining optimum data designs for shadow data sets of manufacturing databases. These subjects will be discussed in details in some future paper.

4. MANUFACTURING INFORMATION'S IMPACT ON COST

Collecting manufacturing information has a cost associated with it, but the cost to the manufacturing process is even greater if pertinent information is not utilized in the appropriate manner. Suppose the test information generated in a manufacturing line starts indicating that the components are not meeting the specifications. The fact that the information is being generated does not guarantee that some corrective measures have been initiated. There can be many reasons for such discrepancies, e.g., (1) there may be a time delay between the instant the information is generated and the initiation of a suitable corrective action (due to cost cutting efforts, people not doing their jobs, etc.); (2) the manufacturer may find it to be more profitable to ignore minor defects (cost less); (3) the manufacturer may not be aware of the appropriate corrective action needed, etc. A simple information utility-cost function is outlined next.

Suppose an item I_j is manufactured at time τ_j, and during manufacturing generates some information for $j = 1, 2, 3, \ldots$. Suppose this information is utilized (i.e., some action is taken) at time t. This information utilization may be as simple as "no action is necessary" to as drastic as "stop the manufacturing line and rework on all items." Sometimes the action may be as harmless as "fix the problem at a later workstation in the manufacturing process." Suppose the information utilization function of the item I_j at time t is denoted by $U(I_j, t, \tau_j)$. Every manufacturing action has associated with it some cost. Analytical methods for determining the relation between utilization of information and cost are difficult. It is also difficult to accurately determine the costs associated with different actions in a manufacturing environment. However, the notion of impact of defect on cost in subsequent stages of manufacturing processes is important. Here we present a simplified linear approach to attacking the problem.

Suppose the cost function associated with the decision to utilize the information generated by the item I_j at time t is denoted by

$$c(U(I_j, t, \tau_j)) \qquad (9)$$

For simplicity we shall decompose the cost function of Eq. (9) into linear components as follows:

$$c(U(I_j, t, \tau_j)) = a_0 + a_1(t - \tau_j) + a_2 u(I_j) + \varepsilon_j \qquad (10)$$

where the ε_j terms are statistical errors and a_0, a_1, and a_2 are constants of the utility-cost function and have to be estimated from actual manufacturing information. These constants also have economic interpretations as follows: a_0 is supposed to reflect the overhead unit cost associated with any action based on manufacturing information; a_1 reflects the cost (per unit time) associated with the delay in initiating an action based on the manufacturing information; a_2 reflects the cost directly related to the action that was taken (e.g., stop the manufacturing process, delay any action until next work station, etc.) in utilizing the information generated by I_j. These are very simple assumptions. It is possible that a_1 may depend on the actual information that was generated (e.g., item slightly out of specification, item way out of specification).

There are many well-known methods of estimating parameters of statistical economic models. Some of these techniques are least-square estimation method, maximum likelihood estimation method, and minimum variance estimation (see Kendall and Stuart, 1961). As these methods are very common they are not repeated in this paper. For estimating a_1 and a_2 from observed data it may be necessary to decompose them into subcomponents that can be assumed to be constant, e.g., a_1 may be decomposed into three components corresponding to (1) the item is less than 5% off from the specification, (2) the item is between 5% and 15% off from specification, (3) the item is more than 15% off from specification. Similarly, a_2 can be decomposed into subcomponents corresponding to different actions for utilization of manufacturing information. If such decompositions are introduced, then the number of parameters in Eq. (10) will be more than three.

The statistical estimates of a_0, a_1, a_2 (based on least-square, maximum likelihood, etc.) will need the following data aggregates:

$$\sum_j c(U(I_j, t, \tau_j)) \quad \sum_j c(U(I_j, t, \tau_j))(t-\tau_j) \quad \sum_j c(U(I_j, t, \tau_j))u(I_j)$$
$$\sum_j (t-\tau_j) \quad \sum_j (t-\tau_j)^2 \quad \sum_j (t-\tau_j)u(I_j)$$
$$\sum_j u(I_j) \quad \sum_j u(I_j)^2 \tag{11}$$

In the relational database structure given in Eq. (5) it is possible to define some of the $\mathrm{AG}(S_i)$ terms to correspond to the aggregates given in Eq. (11), and thus the statistical estimation of a_0, a_1, a_2 can be handled within the framework of the data design discussed in the previous section.

5. CONCLUSIONS

In this paper we have introduced many fundamental database problems associated with statistical process control in an automated manufacturing environment. We have not confined our discussion to classical database management problems suitable for logical processing of information. We have discussed extensively the blending of statistical and logical processing of information in an automated manufacturing environment. We have not discussed the more difficult area of pictorial information processing in manufacturing environment. Like some researchers in statistical process control, we feel that a suitable data model for manufacturing information should contain not only category attributes but also various forms of statistical attributes. We have discussed suitable relational models for representing and storing such information in a database. We have provided many SQL/DB2 statements to achieve these goals. In the context of manufacturing statistics (which can help statistical process control) we have discussed some data structures that can help fast retrieval of statistics needed for real-time manufacturing decisions. We have also introduced the concept of a shadow data set to provide a footprint on the data set associated with the statistical database. It has been established that such shadow data sets have structural similarities with statistical martingale sequences. We have not investigated to a large extent the properties of martingales for manufacturing information processing; much work needs to be done in this area. We have discussed a linear model to predict manufacturing costs based on some key parameters related to the utilization of manufacturing information. These ideas are only the starting point. A considerable amount of theoretical and practical research needs to be done in this new area to achieve long lasting benefits.

ACKNOWLEDGMENT

The authors would like to thank Dr. Kwan Wong of IBM Alamaden Research Center for his support and encouragement of this work.

REFERENCES

Bell Labs. (1982). *Statistical Quality Control Handbook.* Western Electric Co., Indianapolis, Ind.

Box, G. E. P., W. G. Hunter, and J. S. Hunter. (1978). *Statistics for Experimenters.* Wiley, New York.

Codd, E. F. (1970). A relational model of data for large shared data banks. *Communications of the ACM*, *13*(6), 377–387.

Cubitt, R., W. Cooper, and T. Ozsoyoglu. (1986). *Proceedings of the Third International Workshop on Statistical and Scientific Database Management*, Luxembourg, July 22–24.

Ghosh, S. P. (1983). An application of statistical databases in manufacturing testing. *Proceedings of COMPDEC* (*1984*), Vol. 1, pp. 96–103; also published as a IBM Research Report No. RJ 4055.

Ghosh, S. P. (1984). Statistical relational tables for statistical database management. *IEEE Transactions on Software Engineering*, SE-12, No. 12, (1986) pp. 1106–1116. IBM Research Report No. RJ 4394.

Ghosh, S. P. (1985a). Statistics metadata: Linear regression analysis. *Proceedings of International Conference on Foundations of Data Organization*, Kyoto, Japan, May 21–24, pp. 3–12. Also published as IBM Research Report (1984) No. RJ. 4444.

Ghosh, S. P. (1985b). Statistical data reduction for manufacturing testing. IBM Research Report No. 4715. *Proceedings of IEEE International Conference on Data Engineering*, Los Angeles, pp. 58–66.

Ghosh, S. P. (1985c). SIAM: Statistical Information Access Method. IBM Research Report No. 4865. *Proceedings of the Third International Workshop on Statistical and Scientific Database Management*, Luxembourg, July 22–24, 1986, pp. 286–293.

Ghosh, S. P. (1986). *Data Base Organization for Data Management*, 2nd ed. Academic Press, New York.

Hammond, R., and J. L. McCarthy, eds. (1983). *Proceedings of the Second International Workshop on Statistical Database Management*, Los Altos, California, September 27–29.

Johnson, R. R. (1981). Modeling summary data. *Proceedings of ACM SIGMOD International Conference*, pp. 93–97.

Kreps, P. (1982). A semantic core model for statistical and scientific data bases. In *LBL Perspective on Statistical Database Management*, ed. H. K. T. Wong, pp. 129–157. Lawrence Berkeley Laboratory, University of California, Berkeley, Calif.

Kendall, M. G., and A. Stuart. (1958). *The Advanced Theory of Statistics*, Vol. 1, Chapters 1–6. Charles Griffin & Company, London.

Kendall, M. G., and A. Stuart. (1961). *The Advanced Theory of Statistics*, Vol. 2, Chapters 17–20. Hafner Publishing Co., New York.

Loeve, M. (1960). *Probability Theory*, Section 29. Van Nostrand, New York.

Navathe, S., R. Elmasri, and J. Larson. (1986). Integrating users views in database design. *IEEE Computer Magazine*, *19*(1), 50–62.

Rao, C. R. (1952). *Advanced Statistical Methods in Biometric Research.* Wiley, New York.

Roth, M. A., H. F. Korth, and D. S. Batory. (1985). SQL/NF: A query language for non-NF relational databases. University of Texas, Austin, Computer Science publication, TR85-19.

Shoshani, A. (1982). Statistical databases: Characteristics, problems, and some solutions. *Proceedings of the 8th International Conference on Very Large Data Bases* (*VLDB*), pp. 208–222.

SQL/Data System Application Programming. (1983). IBM Program Product. SH24-5018-0.

Su, S. S. Y. W. (1983). SAM*: A semantic association model for corporate and scientific-statistical databases. *Journal of Information Science*, *29*, 151–199.

Su, S. S. Y. W. (1986). Modeling integrated manufacturing data with SAM. *IEEE Computer Magazine*, *19*(1), 34–49.

Wong, H. K. T., ed. (1982a). *A LBL Perspective on Statistical Database Management.* Lawrence Berkeley Laboratory, University of California, Berkeley, Calif.

Wong, H. K. T., ed. (1982b). *Proceedings of the First LBL Workshop on Statistical Database Management*, Menlo Park, Calif., December 2–4, 1981.

SECTION V

Knowledge-Based Systems in Process Control

11
An Expert System Tool for Real-Time Control

K. Kumar Gidwani*
LISP Machine, Inc.
Los Angeles, California

Expert systems are being used to monitor, control, simulate, diagnose, and optimize industrial processes. PICON is a real-time, on-line expert system tool that is ideally suited for applications in which expert decisions are required using live data from a dynamic process source, deep knowledge of process behavior, and heuristic knowledge of the process expert.

This paper provides an overview of the PICON expert system and explains what makes PICON a real-time, on-line expert system. The interactive facility for knowledge-base capture, the real-time interface for data acquisition, simulation capabilities for testing the knowledge base, and operator communication and explanation features provided in the software package are also discussed.

**Current affiliation:* American Express Company, Phoenix, Arizona.

Based on a presentation made at "Statistical Process Control: Keeping Pace with Automated Manufacturing, a National Symposium," sponsored by the Center for Professional Development and the Reliability, Availability and Serviceability Laboratory, College of Engineering and Applied Sciences, Arizona State University, November 6–7, 1986.

1. INTRODUCTION

Artificial intelligence (AI) is the part of computer science concerned with designing intelligent computer systems, that is, systems that exhibit the characteristics we associate with intelligence in human behavior.

There are three identifiable areas of research that are included in discussions of AI:

1. National language processing—investigates methods of allowing the computers to communicate with people in ordinary English. Major applications include database management system interface, intelligent speech processing, and computer interfaces for naive users.
2. Problem solving—discovers solutions to complex problems using heuristics, problem reduction, and other problem solving techniques. Major applications are robot problem solving, object recognition or scene analysis, and intelligent management information systems.
3. Expert systems—computer systems that perform, in very limited domains, at the level of a human expert. Major applications have focussed on medical diagnosis, financial analysis, and industrial automation.

Since the expert systems field promises a great deal of practical application and commercial potential in the near future, it has begun to attract an enormous amount of attention. Expert systems technology is poised to become the first AI technology to have a widespread impact on business and industry.

2. EXPERT SYSTEMS DEFINED

Expert systems are computer programs in which logical reasoning is supplemented by theoretical knowledge, judgment, and rules of thumb. An expert system attempts to simulate human reasoning by chaining together the necessary facts in a given domain to logically arrive at the conclusions.

An expert system differs from a conventional computer program because it stores the expert's knowledge (heuristics,

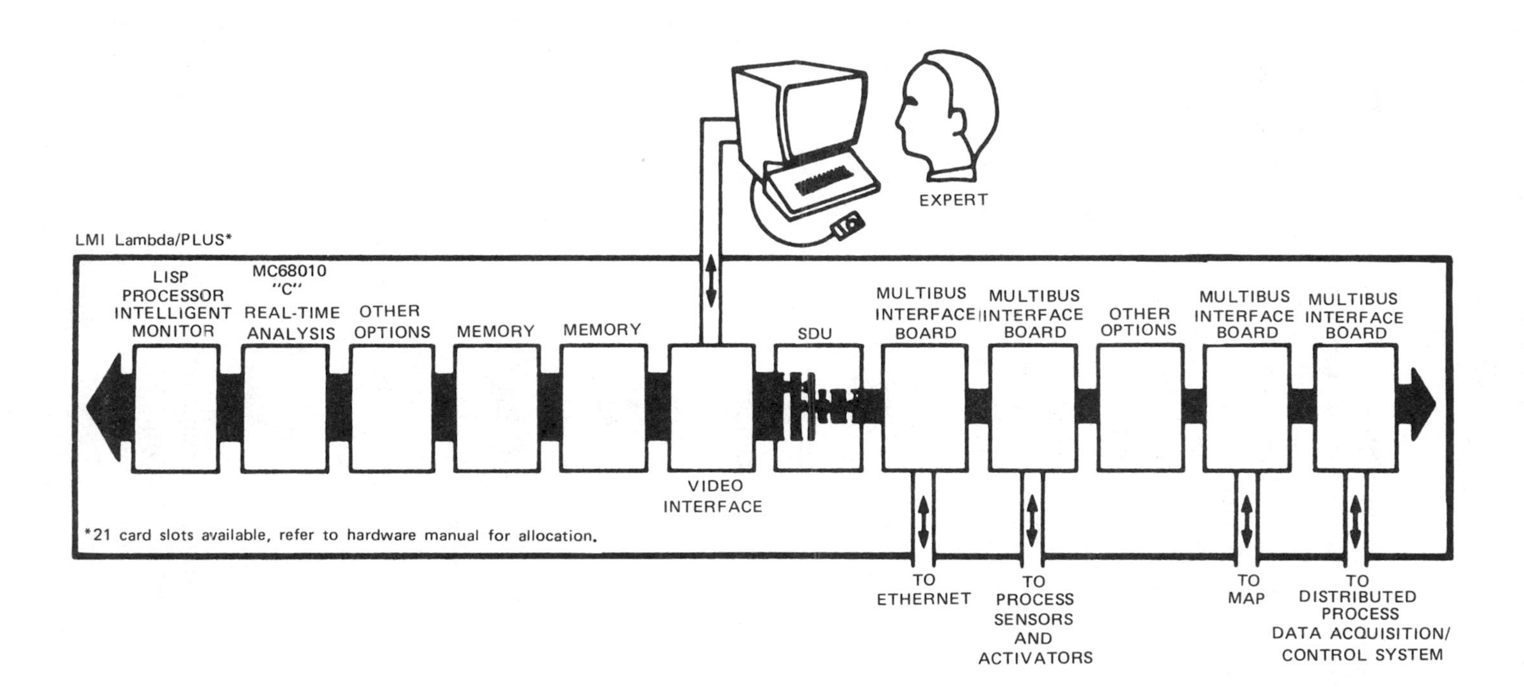

FIGURE 1 Lambda/PLUS typical interfacing options.

models, etc.) apart from the program required to run it. This unique feature allows an expert to concentrate on working with the problem independently of the computer program. An expert system can, in essence, be used by non-computer specialists, making its development and maintenance extremely easy.

The expert system may be interactive and use static data or on-line and use real-time dynamic data. An interactive system requires a user to interact with it to solve a problem or to supply additional information during the execution. On the other hand, an on-line system is capable of communicating with a data acquisition system to access information as required during the decision making. Almost all of the systems built so far are non-real-time interactive systems.

PICON (*process intelligent control*) is the first expert system tool specifically designed for building on-line real-time systems. PICON runs on the LMI Lambda/Plus enhanced LISP machine (Figure 1). The "Plus" is a tightly coupled 68010 co-processor that provides a *real-time intelligent machine environment* (RTIME) under PICON'S control. This architectural breakthrough allows expert systems to be built that can be interfaced with computers or distributed control systems used for industrial process control or programmable controllers, robots, and various types of computer systems used for factory automation.

3. AN OVERVIEW OF PICON

The PICON system is an integrated hardware/software environment that is built using state-of-the-art technology of artificial intelligence. Figure 2 shows the overall structure of the system. The software/hardware implementation of this system requires:

1. An intelligent computer interface to acquire data from sensors and actuators of the physical process as required for making decisions
2. A technique to process incoming data so that it is directly usable for decision making
3. A means of entering the expert's knowledge into the computer's knowledge-base by the domain expert with no AI background

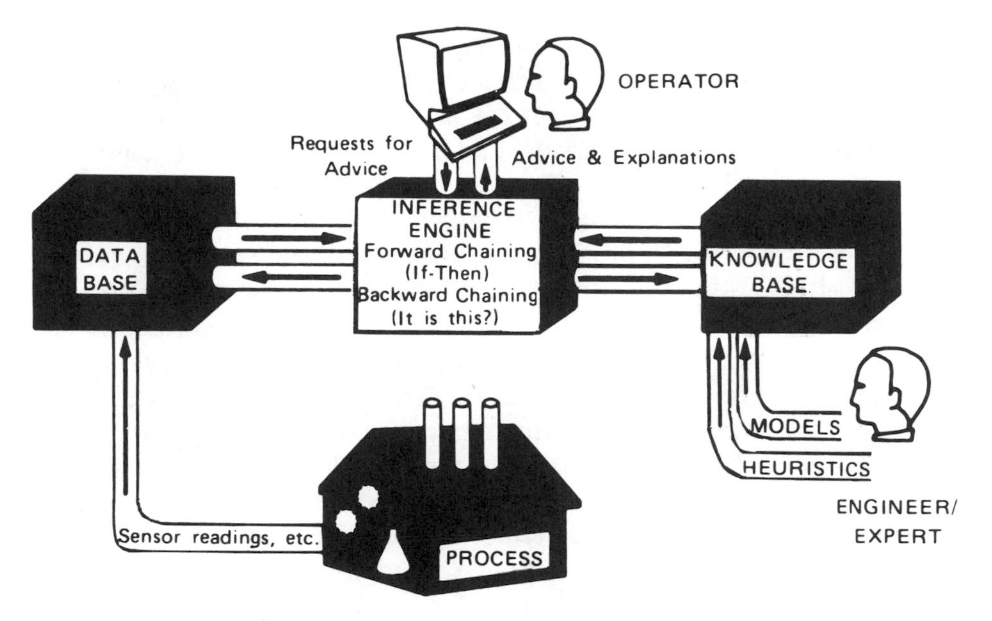

FIGURE 2 Structure of expert system.

4. A means of accessing expert advice and explanations by the process or factory operator
5. An inference engine that is capable of processing large complex problems, applying expert knowledge to real-time data, and responding in a matter of seconds

4. WHAT MAKES PICON A REAL-TIME "ON-LINE" EXPERT SYSTEM?

In order to respond in real time to the thousands of dynamic data points of a large real-time system, PICON was designed with seven unique features.

1. On line data collection: PICON interfaces directly to the process control data highways, factory automation networks, or any other source of dynamic data via its RTIME interface module. It selectively accesses the data needed for inference and decision making. All data is time-stamped and carries a user-selected

currency interval that defines the life of data. Thus, all of PICON'S inferences and actions are based on up-to-date information.

2. Scanning: PICON may, at the user's option, scan certain conditions at regular intervals, looking for incipient upsets, problems, or significant events. The user can specify a scan rate for the rules that control this activity. When invoked, these primary alerting rules cause PICON to focus, as discussed below. This technique allows PICON to deal with very large problems and still deliver the real-time response suitable for the most stringent application.

3. Focus: Some sensors need to be read and some rules tested only when a certain situation has occurred. These secondary rule-frames are activated by the primary alerting rules, causing PICON to focus its attention on parts of the process related to the actual or developing problem detected by the primary rules. The secondary rules can activate even more rules through forward-chaining (if–then), which can proceed with as much in-depth search as required. To check necessary conditions for testing a hypothesis, it backward-chains to initiate data collection for the needed sensors. This technique, unique to PICON, is extremely powerful when a system must deliver fast on-line responses, for example, in a nuclear power plant where the knowledge base and data source can be very large and dynamic.

4. Scheduling: At the heart of the PICON inference engine is the scheduler. On-line applications require that data be obtained from other control or automation systems, but it may not arrive in time for a particular inference. PICON's scheduler is responsible for interrupting and resuming inferences and actions that are not complete because of lack of data; for taking alternative actions when an inference does not complete in a reasonable amount of time; and for keeping many activities going without ambiguity. It can schedule any number of activities, such as testing a rule to occur on a regular, cyclic basis. It can also schedule any activity to occur at some specific time in the future, whether 3 seconds or 3 weeks.

5. Long-term strategy: An expert system that could respond only to a current situation would be of limited utility in most applications. PICON is designed to deal effectively with the rates of change, and any abnormal trends in the process. It also keeps track of its own plans instead of working with a fixed algorithm (single

strategy). PICON can switch strategies or selectively invoke relevant techniques applicable at a given time.

6. Background maintenance: Because of the large memory of the LMI Lambda/Plus LISP machine in which PICON runs, and PICON's unique scheduling facility, many thousands of activities can be scheduled in the background without interfering with PICON's ability to focus on the current situation. These may include routine inspection of sensors or other pieces of equipment to detect failures or marginal performance. Thus, PICON is capable of managing a very large process in real time.

7. Simulation: PICON is supplied with a dynamic simulator that has the same user interface as the knowledge-base editor. The simulator is a distinct module that supplies sensor values to the inference engine as though they were obtained from a real plant. The user can select the source, i.e., simulation or real process from which PICON gets its data. The simulation can be developed and tested incrementally as the knowledge base is being built, making it an ideal tool for testing the knowledge base and checking PICON's response to both normal and abnormal conditions. The simulator is also very useful for training operators, as it can be used to expose them to situations seldom encountered.

5. INTERACTIVE FACILITY FOR KNOWLEDGE-BASE CAPTURE

PICON is designed for use by application domain experts with no previous background in artificial intelligence. It provides easy-to-use facilities for entering process engineering knowledge and operator's heuristics into the knowledge base. Two types of tools are available to capture knowledge: a schematic capture tool for process knowledge, and a structured natural-language interface for heuristic knowledge.

The domain expert builds a schematic of the plant or a factory by selecting ICONs to represent sensors and other plant equipment and connections to represent piping or instrumentation that links ICONs together (Figure 3). The inquiries then proceed, by pop-up windows, to request information required for the knowledge base,

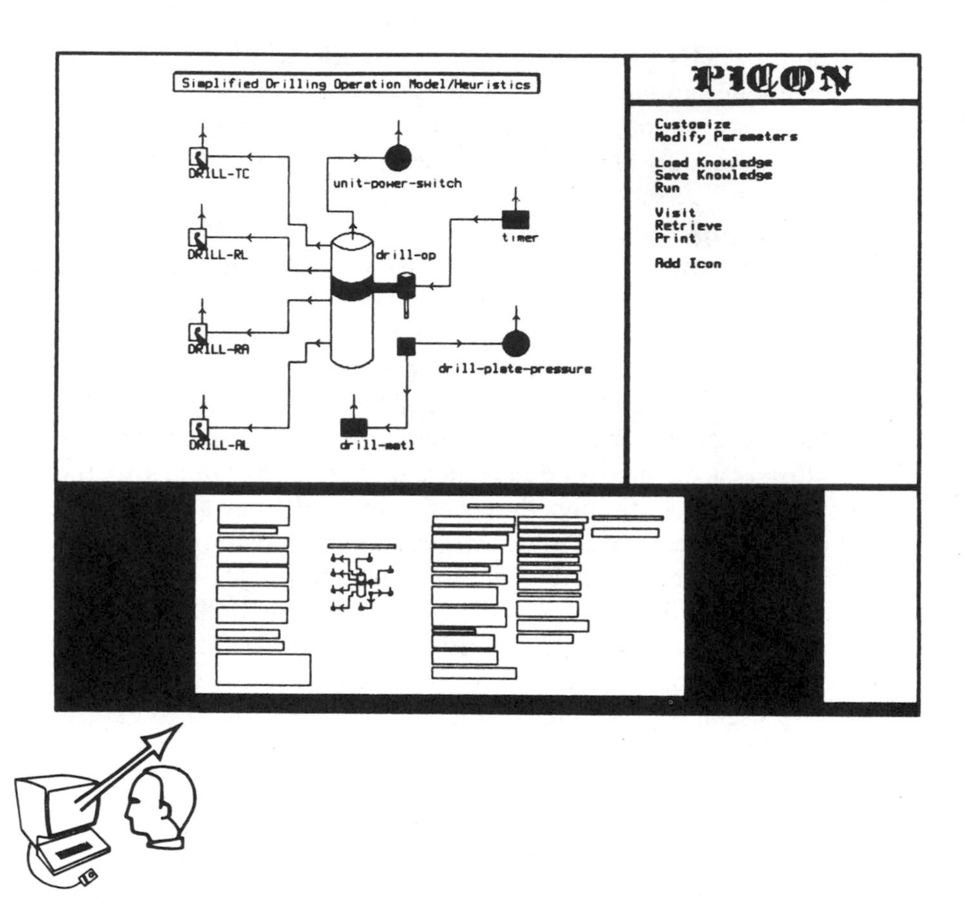

FIGURE 3 Part of a simplified process schematic.

such as which of the various types of an ICON is being specified, and specific parameter values.

ICONs are organized into frames. Frames, in turn, contain slots in which knowledge, such as ICON attributes, can be stored (Figure 4). Each ICON is a specific instance of an ICON type.

By moving the mouse to the tag name for the sensor in the attribute list and clicking a button, the user can enter the tag name. This establishes access to the measurement of the control system.

Similarly, the user can enter the complete process knowledge base along with process parameters in the corresponding attributes

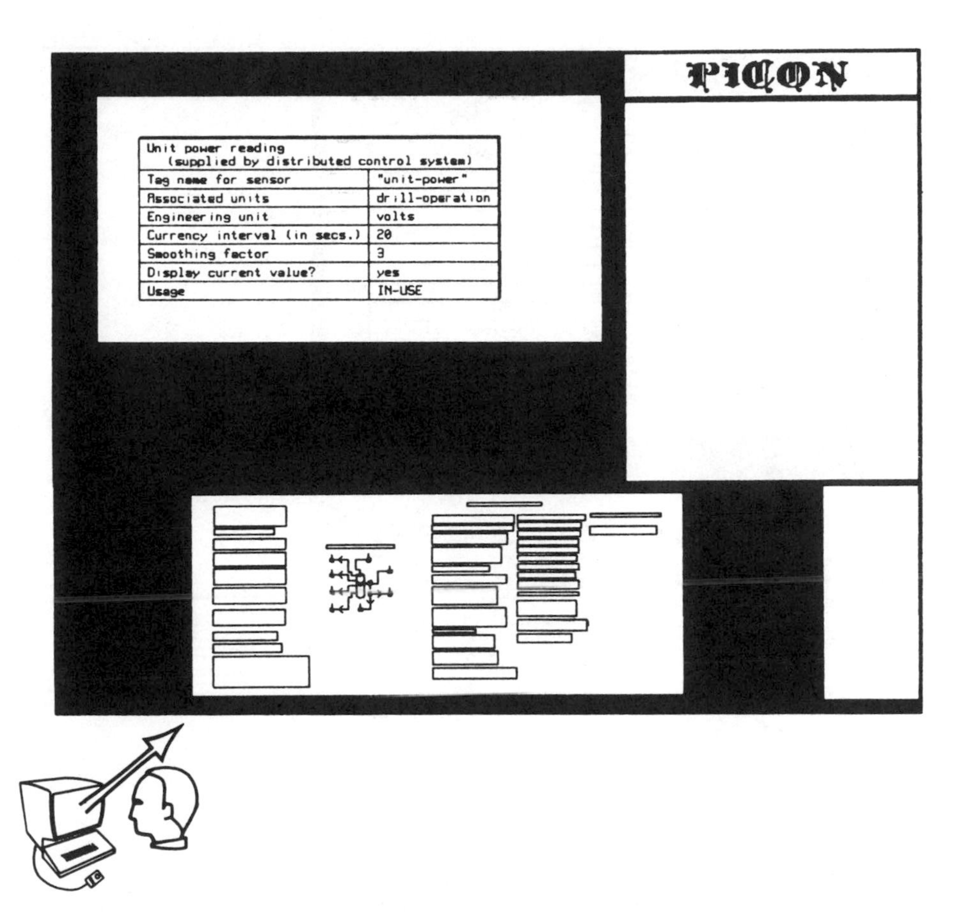

Unit power reading (supplied by distributed control system)	
Tag name for sensor	"unit-power"
Associated units	drill-operation
Engineering unit	volts
Currency interval (in secs.)	20
Smoothing factor	3
Display current value?	yes
Usage	IN-USE

FIGURE 4 Abbreviated frame for an object.

lists of ICONs. The schematic can be viewed in multiple windows, at a variety of scales, which allows conceptualization and representation of a large process.

Rules, like ICONs, are organized in frames (Figure 5) along with their attributes lists. The rules are added by means of keyboard commands and menu selections, which allow the user to create the rule blocks.

Additional tools provided to facilitate maintenance of a knowledge base are the "VISIT" and "RETRIEVE" features. These allow one to search the entire set of rules to check for consistency or the

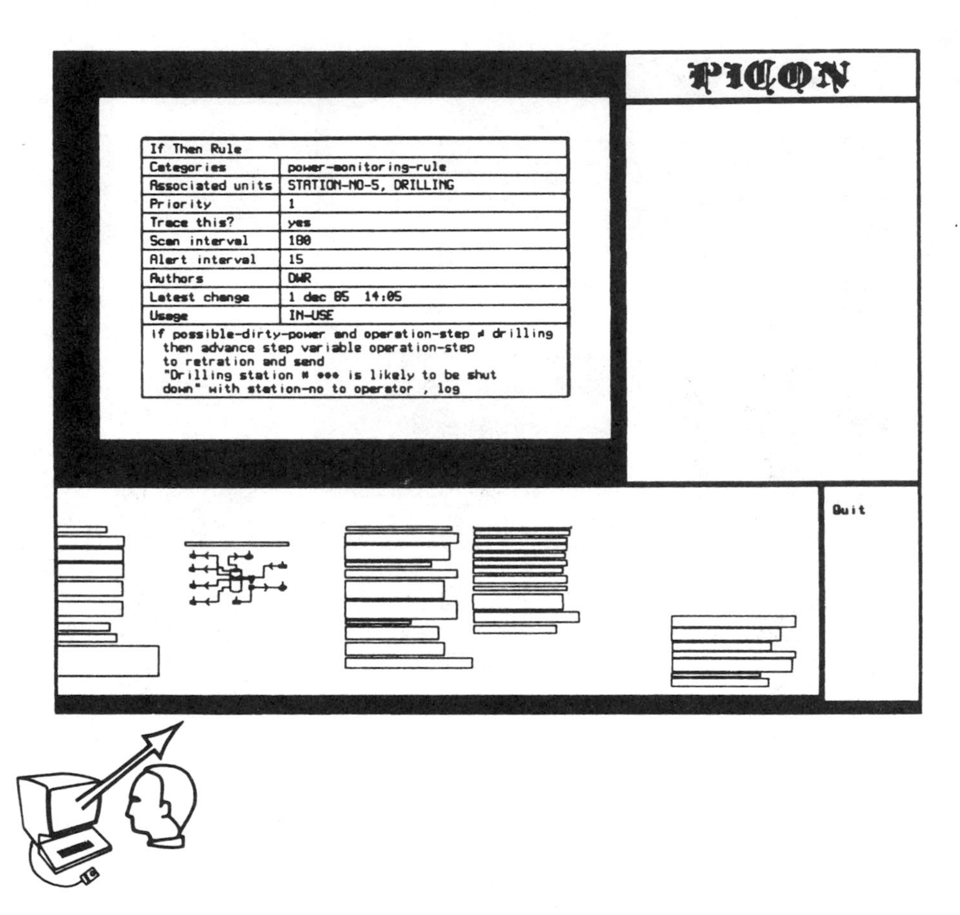

FIGURE 5 Example of an abbreviated rule frame.

schematic integrity by specifying certain descriptive information such as a character string or author of a rule.

6. REAL-TIME INTERFACE FOR DATA ACQUISITION

PICON interfaces with the real-time source of data such as a process control system for a real-time data acquisition via the

RTIME, real-time interface. The RTIME module of PICON is written in "C" language and runs on the MC68010 coprocessor in the multiprocessor PICON system. RTIME consists of three functional submodules:

1. The LISP communicator module: This module provides efficient and effective communication between the programs that run in the LISP and RTIME processors. All LISP communications, including shared memory used for storage and access of data in engineering units, are transparent to the user.
2. The execution module: In this module, RTIME performs all of the preprocessing functions called by the user as a part of the node descriptor table. A node is a designated data source in a process or a plant network. RTIME is enriched with a library of commonly used algorithms. Specialized algorithms can be added to this library to suit a problem domain.
3. The I/O driver: This part of RTIME inputs and outputs data on the interconnect device chosen to communicate with the external system. Standard communications are via Multibus, high-speed RS232, Ethernet, and other Multibus-compatible interfaces. Since PICON is applicable to a variety of data-acquisition systems with differing protocols, it is often necessary to customize this part of RTIME to a specific network.

7. OPERATOR COMMUNICATION AND EXPLANATION FACILITY

PICON can send expert advice to the operator and/or to the process to change a process variable that is, controller setpoint in a closed-loop situation. The system is designed to be compatible with existing color displays commonly offered by process control vendors to display PICON decisions on these terminals. An example is shown in Figure 6.

The explanation facility in the PICON allows the user to ask the system to explain its decision process. This is presented to the user graphically in the form of a decision tree. The decision tree is dynamically created by the system for every situation. Modifica-

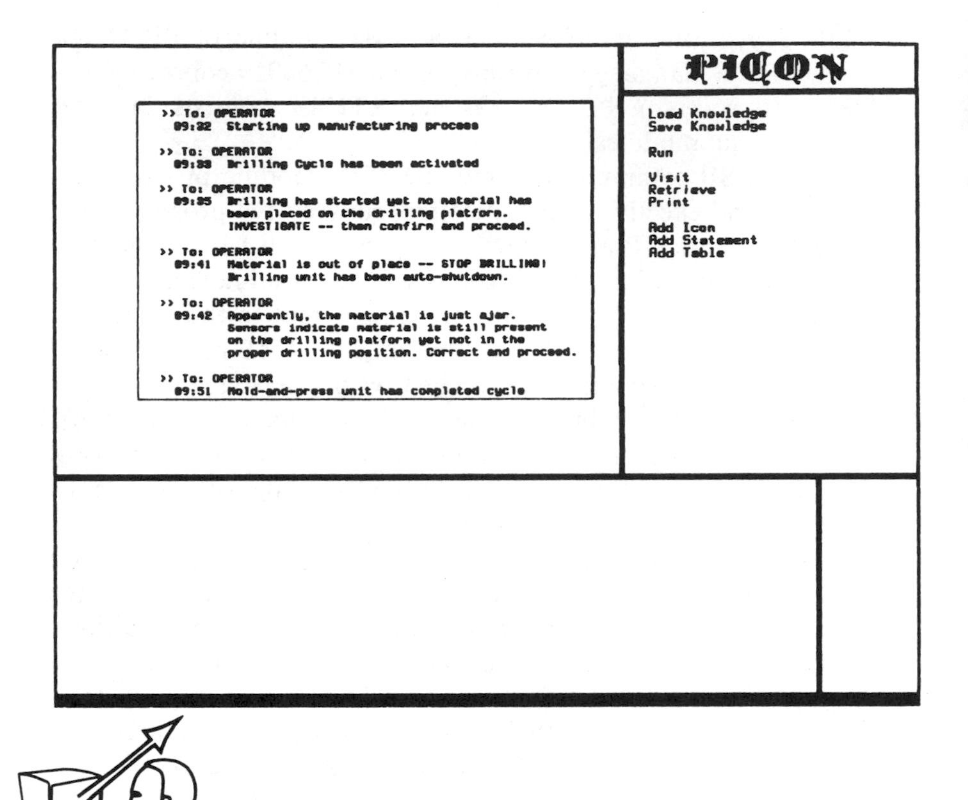

FIGURE 6 An example of PICON's operator messages.

tions and additions to the knowledge base can be made without stopping PICON. Figure 7 shows the explanation facility for a simplified example. The complete decision tree is shown in the bottom box. Each of the blocks in the tree could be either a rule, a sensor value, or a computed variable. The upper box is used to display an expanded picture of one of the blocks from the decision tree.

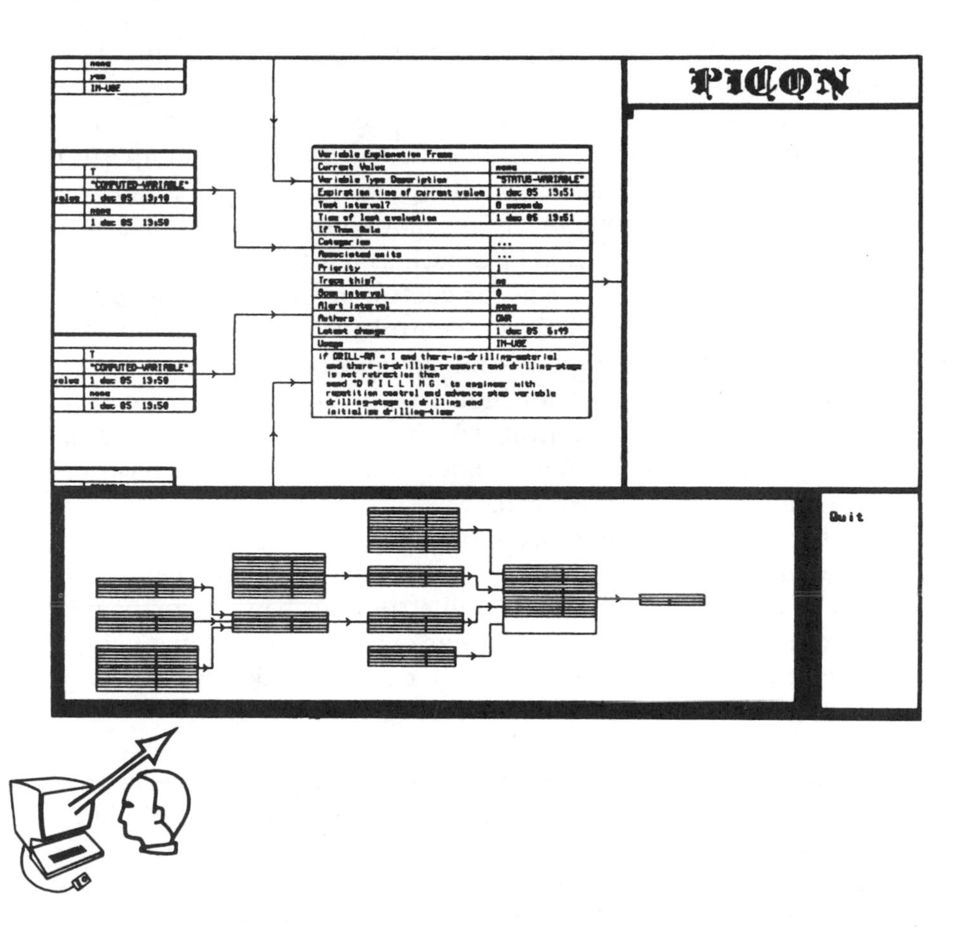

FIGURE 7 Display showing decision tree.

8. EXPERT SYSTEMS APPLICATIONS

The most obvious application is during a major plant upset, where an operator must cope with hundreds of alarms in a limited time. In the case of a nuclear power plant, for example, the operator may have 800 alarms in the first 2 minutes of a major upset. An operator may be an expert, but in a short time it is difficult for any human operator to diagnose multiple problems involving such quantities of information.

An expert system could also be applied to detect measurement integrity or process integrity problems. An elementary example is to calculate a material balance around a process and to compare it with the measured flows. A discrepancy could indicate a measurement failure or a process failure, such as a leak.

The use of expert systems for control has been investigated by Astrom and Anton (1984). Expert control offers a compromise when accurate and tractable models for optimization are not available. This is the common case for industrial processes, and the current alternative is manual control, which may be subject to varying quality. Expert control offers the potential for providing consistent expert advice or direct control.

Expert systems for process startup and shutdown are a particular area of interest. These operations are done infrequently, but require considerable expertise, and errors can be very costly. One petrochemical company described an experience where an operator mistake during a shutdown of a unit caused a shutdown of a large part of the refinery operations for more than a day.

The "teachable" nature of expert systems opens new opportunities for plant control approaches. Expert systems have the characteristic that the knowledge base can be augmented and modified. This allows adaptation to changes in plant condition and operating objectives—the equivalent of changes in operating instructions, which is a common occurrence in plant operations. An expert system, once the knowledge is defined, could serve as a training vehicle for new operators. In many cases expertise is built up over many years of operation, and is lost when an employee leaves or retires. The knowledge base of an expert system can be expected to become a corporate resource, available without disruption for training, simulation, and other purposes as well as being used for advice and control for plant operations.

REFERENCE

Astrom, K. J., and J. J. Anton. (1984). Expert control. *IFAC World Congress*, Budapest.

12
Sensor Fusion

Steven R. LeClair
U.S. Air Force
Wright-Patterson Air Force Base, Ohio
Jack Park*
ThinkAlong Software Inc.
Brownsville, California

1. SENSOR FUSION

This chapter describes a control system architecture based on heuristics as opposed to mathematical models. More specifically, the subsequent discussion presents a design for process control that is predicated on "qualitative" reasoning about physical systems (Bobrow et al., 1985).

1.1 Introduction

In most, if not all, process control problems there is a need to collect and understand aggregate (fused) data from multiple sources (sensors). The concept of *sensor fusion* is analogous to human sensory processing where the "qualitative" evaluation of combined

**Current affiliation:* U.S. Air Force, Wright-Patterson Air Force Base, Ohio.

input from several senses produces a richer and more reliable perception of the environment than does evaluation of a single sense or separate evaluation of multiple-sense data (Garrett, Lee, & LeClair, 1987).

To accomplish qualitative evaluation of fused sensor data we must take advantage of symbolic processing afforded by *artificial intelligence* (AI) technology. When considering the advantages afforded by AI techniques, the most significant benefits are those associated with complementing and improving human decision-making activities. Bullers et al. (1980, pp. 351–352) noted that

> While computer technology can rapidly process large amounts of data by sophisticated logic, many of the necessary decisions must wait until human operators can sift through the data, become familiar with the system status, and select proper actions. This is where artificial intelligence techniques can provide better planning and control.... AI techniques can be used to develop decision aids that are capable of handling large streams of data as well as performing logic manipulation for conflict resolution, sequencing and resource allocation, etc.

Figure 1 provides some perspective regarding the evolution of planning, scheduling, and control and the expected contribution from the application of AI techniques. The *adaptive control* example illustrates the state-of-the-art today, where (without AI) the control activity must rely on feedback. All is well as long as the environment is relatively stable and the priorities are fixed—as the prescribed knowledge will adapt accordingly. But once the "requisite variety" or the breadth of control is exceeded or the priorities are changed, the prescribed knowledge becomes inadequate and thus the ability to fuse multiple-sense data and learn new knowledge becomes necessary. It is envisioned that AI will enable this next step in sophistication such that a machine can perform these activities much like a human, i.e., continually learning how to improve performance.

Although the sensor fusion system to be described in the following paragraphs is a generic tool and thereby adaptable to any decision-making activity, the specific problem of "process control" of curing graphite–epoxy laminates will be used to illustrate the operation of system components (LeClair, et al, 1987).

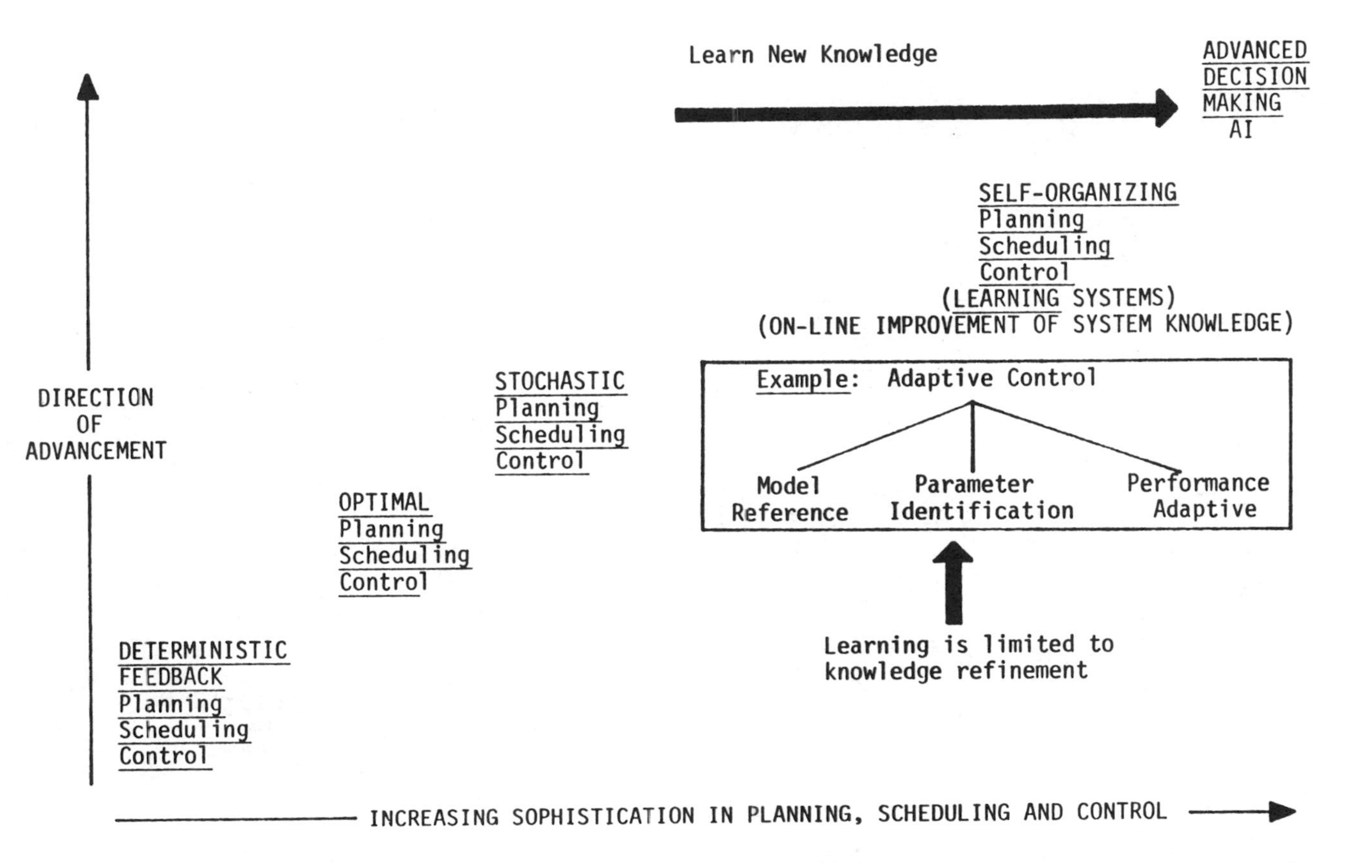

FIGURE 1 Increasing sophistication on the path to intelligent planning, scheduling, and control.

1.2 Purpose

Figure 2 illustrates a complex process: the fabrication and assembly of a physical structure. This process requires the completion of many subprocesses, each representable as a goal. In order that each goal may be achieved, it is assumed that *control* is required. Also of importance during any process is the possibility that conditions will arise that deviate from the expected process history. In order to properly deal with these exceptions (faults), fault diagnosis is also required.

The universe of process control, depicted in Figure 3, illustrates the complex relationship that exists between natural processes (e.g., empirical laws and theories such as thermodynamics, quantum mechanics, statics, economics, etc.) and synthetic processes—those contrived by humans to make use of natural processes to achieve some goal(s). One view of a system contrived to achieve some goal(s) in this context is as an *experiment*. Any experiment is an interaction between some desired synthetic process and the natural

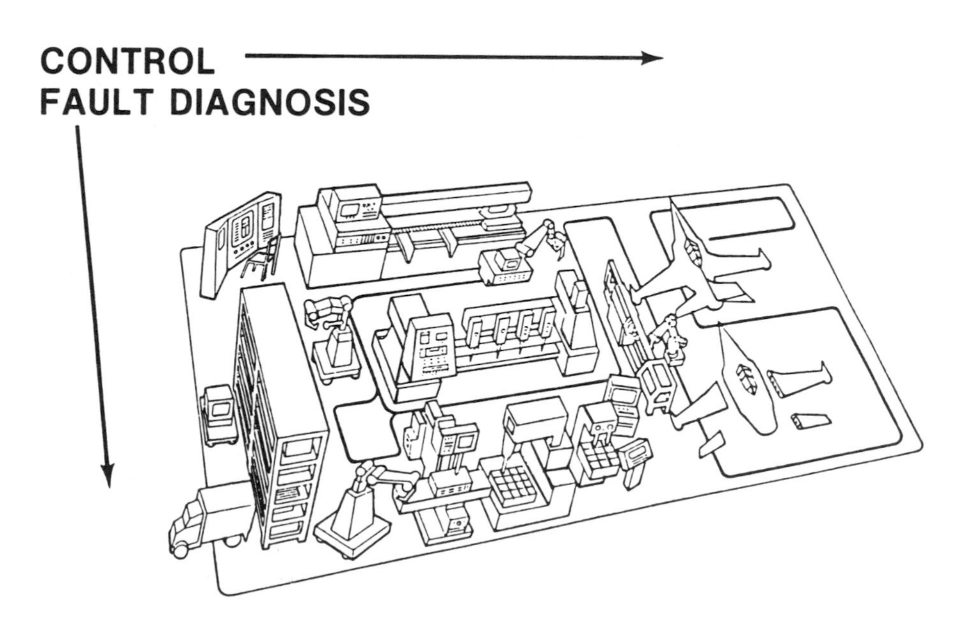

FIGURE 2 Sensor fusion.

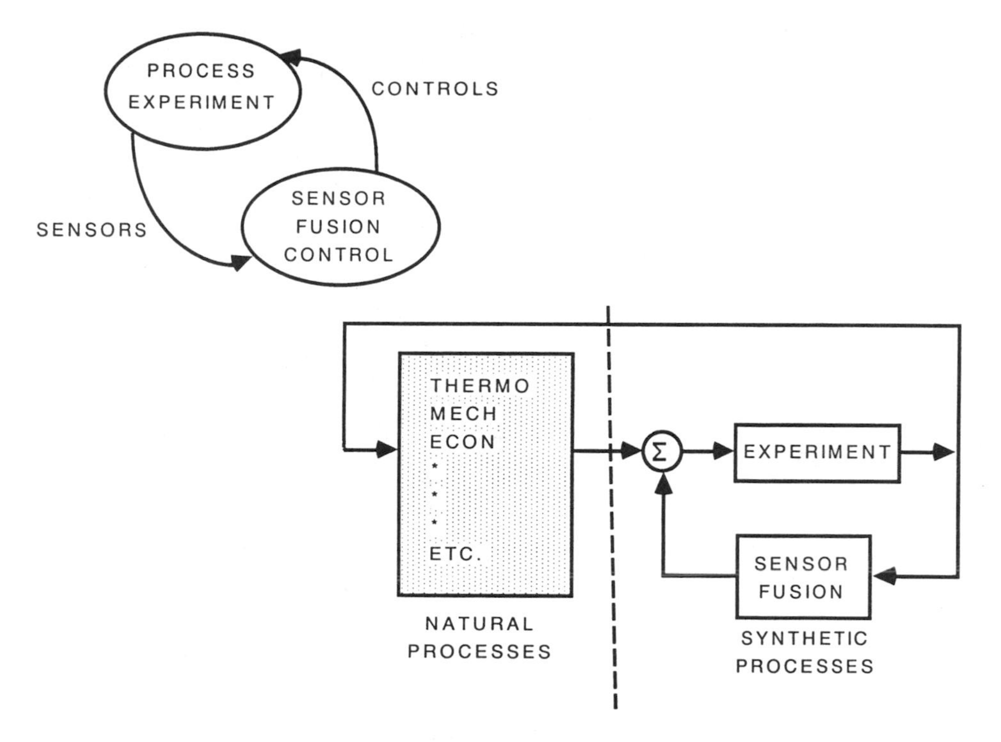

FIGURE 3 Process control.

processes that affect it. Thus, the purpose of a *sensor fusion system* is the mediation of effects of natural processes on an experiment in order to achieve some desired goal(s).

1.3 Architecture

The sensor fusion system architecture includes both functional and hierarchial aspects. First we introduce the functional aspects of sensor fusion; then we introduce the hierarchial aspects. Figure 4 illustrates a system capable of *sensing* or *observing* the experiment as it progresses, determining the meaning(s) of the sensed data, making decisions, and issuing commands that guide or otherwise affect the experiment. This data flow, from sensors to commands, is the functional architecture upon which sensor fusion is designed.

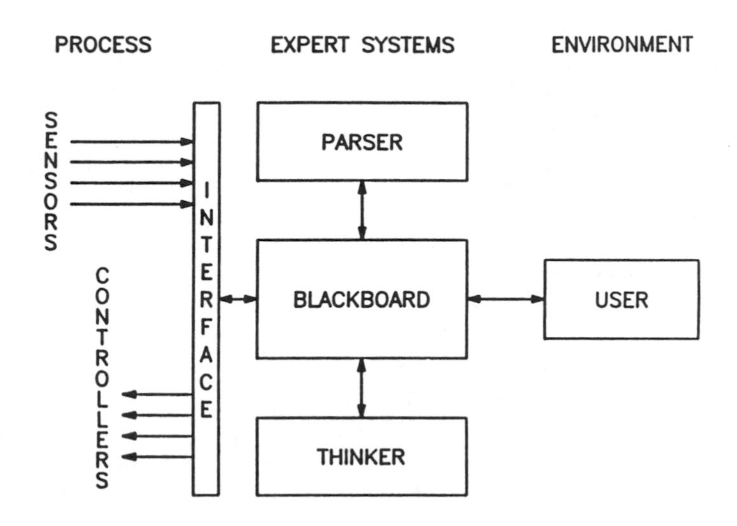

FIGURE 4 Sensor fusion information flow.

Based on this data flow, the control cycle iterates between two constituent tasks:

1. Recognizing the situation (parsing)
2. Responding with commands (thinking)

1.3.1 Parsing

In order to understand the state of a process, the continuous flow of information is transformed into state information about the environment. This transformation is called *parsing*, and makes use of a metaphor: the natural language of any process (experiment) is the flow of state data through sensors. By applying this metaphor, we are able to build a parser capable of the necessary transformations. As a simple linguistic example of parsing, consider the sentence,

John picked up the hammer.

This sentence is stated in a natural language, and has a meaning capable of being found by a parser. It is not a simple sentence since the word "picked" needs disambiguation of its several potential meanings: *grasp* and *choose* are two such candidate meanings. A simple parser would notice the next word in the sentence "up" and use its position to disambiguate the word "picked," where "picked up" parses to the primitive process of grasping. Thus, a simplified computer memory structure that represents an instance of the primitive grasp process may be built from this sentence:

```
GRASP-1       (an instance of the grasp process)
  ACTOR       (an independent player in the process)
    JOHN
      TYPE human      (facts about individuals)
      AGE etc.
  OBJECT      (a dependent player in the process)
    HAMMER
      TYPE tool
      SIZE etc.
  INSTRUMENT      (other related information)
    FINGERS
      PART-OF John
```

With this memory structure, the computer is capable of answering questions about the process, e.g., the parser transforms sentences into memory structures with answers to questions like:

What changed?
Who (what) caused the change?

In the sensor fusion system, the parser transforms sensed data into similar memory structures, finding candidate *process-instances* to explain the data. Using the simple sentence example above, the data flowing to the parser is the sentence. This data is transformed into the memory elements representing the instance of grasp: grasp-1.

1.3.2 Thinking

After a situation is understood and mapped to a memory structure, some process decisions may be necessary. Since all decisions are ordered, only those decisions that are "in-order" are executed through commands issued by the thinker.

Using the simple sentence example above, the thinker would

answer such questions as:

What was picked up?
Who picked up the hammer?
Was my goal to pick up the hammer achieved?

These examples of the "question answering" process lead to control questions regarding the experiment. Such questions include:

Is the process progressing as expected?
What is the next appropriate command?

1.3.3 The Blackboard

The blackboard model of problem solving is a highly structured special case of opportunistic problem solving—the application of either backward or forward reasoning at the most opportune time. The model is usually described as consisting of three major components: (1) the knowledge sources (which for sensor fusion are currently the parser and thinker), (2) the blackboard data structure to enable communication and interaction among the knowledge sources, and (3) control of knowledge source responses to changes in the blackboard (Ni, 1986).

The following analogy illustrates how the sensor fusion system uses a blackboard. Imagine two different scientists working on the same process; one has the task of figuring out the situation in an ongoing experiment, and the other has the task of deciding what to do next (these are the parser and thinker entities of sensor fusion).

Now, imagine that these two scientists have a difficult time communicating because they think at different rates, and perhaps, in different languages. However, they do need to communicate, for if one scientist does not know what the other has found, progress is impossible.

Then imagine that we give the two scientists a uniform method of communicating: a blackboard in a separate room. The first scientist walks into the room and posts what is presently understood about the situation, then walks out. The second scientist walks into the room, reads what is understood about the situation, and posts what should be done about it.

Finally, a blackboard monitor walks into the room and "tidies up." This monitor collects all commands issued and executes them. Such commands include calling for more data from the experiment, writing progress reports for a "peer review" committee (the process operators), filing information in a history file on the project, or retrieving historical information.

1.3.4 The Software

Organization of the sensor fusion software is depicted in Figure 5. As currently configured this organization consists of two separate programs built on top of a core inference system called Expert 5. The two programs are layered so that higher forms of "knowledge" about processes may draw on more primitive knowledge deeper in the program structure. The two programs are:

1. The sensor fusion shell
2. The user's knowledge base program

The sensor fusion shell provides a useful library of behaviors (turning on or off a heater, reading a sensor, etc.) and a library of primitive processes, such as heat flow—heating or cooling, and state changes—freezing, melting, boiling.

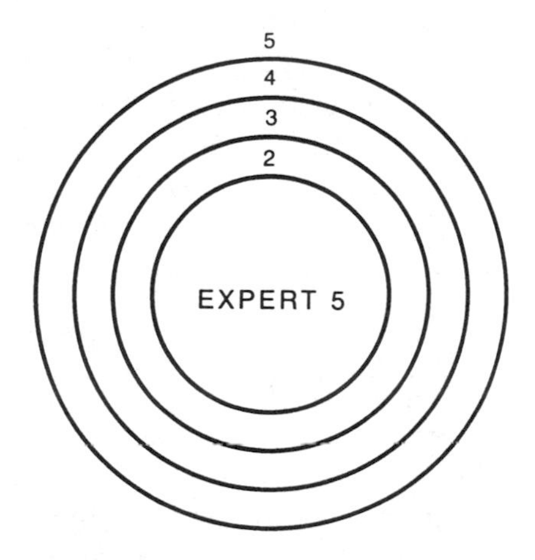

FIGURE 5 Sensor fusion organization.

The user's knowledge base program uses this library of processes and behaviors to describe the experiment and its goals. This description of the experiment does *not* include step-by-step instructions for conducting the experiment; the description only includes a list of processes, and the goals and history of the process. Thus, the sensor fusion system offers a "declarative" or nonprescriptive method for describing an experiment, which differs from the traditional procedural approach in that the order of the process steps are not specified. As an example, a step-by-step set of instructions for baking a cake might include:

1. Preheat oven to 350 degrees.
2. Bake for 20 minutes.

This set of instructions does not include the possibility that the complex chemical reactions and physical processes involved in the transformation of a batch of dough to a cake (the baking process) may not occur as expected under the environmental conditions and specific ingredients used.

A different approach is to *declare* the constraints on the baking process, and initiate the sensor fusion system to conduct the exercise:

1. Maximum temperature = 350
2. Done when cake "springiness" = some value

Here, we introduce a notion any baker already uses: even though the instructions originally read to bake for a certain time, most bakers quit when the cake springs back when poked. So, we invent some springiness value, locate a sensor capable of monitoring that value, declare the limits of the process, and establish a process instance in the sensor fusion system process library that uses the primitive behavior "turn up the heat" to control this process.

2. QP-THEORY OVERVIEW

2.1 Introduction

The sensor fusion system draws together a functional architecture designed for process control and a hierarchial architecture designed

EVERY PROCESS HAS:

- ACTORS
  THE INDIVIDUALS IN THE PROCESS
- PRECONDITIONS
  TIME / STATE REQUIREMENTS BEFORE ACTIVE
- EFFECTS
  TIME / STATE CONDITIONS AFTER ACTIVE
- PARAMETERS
  STRUCTURE AND QUANTITIES

FIGURE 6 Qualitative process theory (QP-theory).

to ease the burden of describing (to the computer) the process to be controlled. The underlying paradigm selected for this architecture is *qualitative process theory* (QP-theory).

The basic notion of QP-theory is that of a *process*. Figure 6 illustrates the concept of a process description. A process is a collection of actions (primitive behaviors) designed to bring about some change. This change may be the result of many partial changes, or it may be a single change. A process instance describes a specific "partial" change. Thus, the process of curing laminated graphite–epoxy composites in an autoclave involves numerous process instances each of which describes some portion of the total change involved.

Types of changes that occur in processes are (1) spatial and (2) temporal.

Spatial changes are those that involve the physical world: changes in position (due to motion), changes in structure (due to physical or chemical processes), and the like. Temporal changes are those that involve time; all process instances occur in time windows variously described as moments, instants, and the like. Waiting is a process that involves (at a surface level) only a change in time.

2.2 Process Description

Each process (or process instance) is a description of objects, their properties, and their relations with each other. Since we are dealing at a fundamental level of knowledge, we must be specific in the description of objects, their properties, and their relations that may seem "common sense" to us but must be spelled out for a computer.

Many of the fundamental concepts of qualitative reasoning involve the notion of *common sense*. This is so because often we tend to overlook the obvious when we define a knowledge base. Unfortunately, a computer is not blessed with a "built-in" body of commonsense knowledge; that body of knowledge must be explicitly provided. In the parsing example, just about everybody knows a person grasps a hammer with fingers; the computer needs to be told this explicitly. Thus, much of the sensor fusion system either predescribes common sense issues of process control, or provides a rich body of tools for the user to add further common sense to a knowledge base.

2.2.1 Objects

The objects in process descriptions are generally matter. From physics, we know that properties of matter are:

Has mass
Occupies space
exists in one of three states: solid, liquid, gas.

For some process descriptions, it will not be necessary to deal with all that is known about objects. For example, one kind of process is a transaction—say, a monetary transaction such as a purchase or sale. In such a case, we need not deal with the fact that money has mass, and is generally a solid object; we need only describe money as an object involved in the transaction, and that its primary attribute is its dollar value. This process does not, for most purposes, need to consider the energy involved in the transaction, although the dollar value, in one point of view, may be regarded as an "energy level"—the capacity of the transaction to do work of one sort or another.

2.2.2 *Relations*

Objects are related to each other in a process. The notion of a relation stems solely from the nature of the *change* involved in the process. For example, a quantity of liquid is changed to a quantity of ice (a solid) in a freezing process. The objects are a thermal energy source (the liquid) and a thermal energy sink or destination. The major relation this process describes is the existence of conditions that enable the flow of heat energy from the liquid object to the thermal energy sink. This heat flow is dependent upon two conditions: the temperature of the thermal energy sink is below that of the liquid, and a path exists by which heat may flow between the liquid and the heat sink.

Freezing is a bounded process—it may be one of many process instances in a much larger process (e.g., the weather). As a bounded process, it starts when the preconditions are met, and ends with the change of state from liquid to solid.

3. QP KNOWLEDGE BASE INTRODUCTORY EXAMPLE

3.1 Introduction

This section provides a first example of a process description for the purpose of introducing the necessary terms:

Objects—actors in a process
Preconditions for a process to occur
Effects of a process
Parameters involved in a process

Further, this example introduces the notion of *parsing* to understand what the states of the environment mean (e.g., which process-instance is currently active). Also introduced is the notion of *thinking* about the currently active process-instance (e.g., which states and quantities change, what to look for/to expect next).

3.2 A Transaction Process Example

A process is a sequence of events that results in some change in the environment. An example process is a monetary transaction. For this discussion, the transaction will be called BUY. In this process, possession of some money is exchanged for possession of some other commodity (e.g., food.)

In keeping with our description of a QP knowledge base, we can define the various parameters as follows:

Process: BUY (a transaction process)
- The ACTORS in this process are:
  - A buyer
  - A seller
- The PRECONDITIONS are:
  - The buyer has a sufficient amount of money available at the time of the transaction.
  - The seller has the commodity to be sold at a price at the time of the transaction.
  - The buyer wants to make the purchase.
  - The seller wants to make the sale.
  - Both actors agree on the terms.
- The EFFECTS of the process occurring are:
  - The buyer will no longer have possession of the amount of money transferred.
  - The seller will have new possession of the amount of money transferred.
  - The buyer will have new possession of the commodity purchased.
  - The seller will no longer have possession of the commidity sold.
- The PARAMETERS involved in the BUY process are:
  - Money
    - An amount
    - In possession of the buyer
  - A commodity for sale
    - In possession of the seller

This BUY transaction fully details the various parameters involved in the BUY process. With this example, we can view the functions of the parser and the thinker in the sensor fusion architecture.

This particular example makes clear the issue of common sense in a knowledge base. To most of us, the idea that a buyer hands money to a seller in exchange for goods or services is generally taken for granted. A knowledge base built for the sensor fusion system must not allow such obvious factors to be taken for granted; explicit listing of all factors involved in the definition of a particular process is required.

3.3 Parsing the Example Process

The parser is tasked to look at the preconditions and determine if the BUY process is active. This amounts to comparison of the motives of the actors (i.e., are the indicated values appropriate—does the buyer want to buy and the seller want to sell?), comparison of the terms—price and amount of cash available, and a determination if the terms are agreeable (an optional check). If all terms of the preconditions are met, the parser marks this process as active and passes it to the blackboard, where the thinker accesses it.

3.4 Thinking About the Example Process

The Thinker is tasked to confirm the results of the parse and to respond by dealing with the process effects. If the parse is valid, dealing with the process effects means marking the possession relations in the data base to reflect new values. At the end of this activity, the database would reflect a lesser amount of money available to the buyer, and greater amount of money available to the seller, and so on.

4. QP KNOWLEDGE BASE DESIGN

4.1 Introduction

This section outlines the steps to follow in building a sensor fusion knowledge base. In order to fully understand this activity, it is useful to understand the system architecture, which is discussed next.

4.2 Sensor Fusion Architecture

A sensor fusion knowledge base is, in fact, three separate knowledge bases integrated into one. Recall (Figure 4, sensor fusion system block diagram) that there are three separate functional components to a sensor fusion system:

1. A blackboard monitor
2. A parser
3. A thinker

Each of these three components is a separate expert system with its own knowledge base. The tasks of each component are:

1. Blackboard (BB) monitor
   - Retrieves data from sensors and posts to parser BB
   - Interprets postings of thinker and:
     - Issues commands
     - Types screen messages
     - Obtains data from keyboard to post on BB as needed
     - Swaps parser and thinker BB areas
   - Moves (with swap) BBs back to thinker and parser
   - Keeps track of time
2. Parser
   - Reads data from BB (sensor raw data)
   - Uses domain knowledge, plus postings from last thinker cycle to parse sensed data into useable form, e.g., finds the process-instance that applies to present data
   - Posts to BB results of parse
   - Posts to BB any parse failure encountered
   - Waits for next cycle
3. Thinker
   - Reads from BB results of parse
   - Uses parse result, plus history, plus present data to verify parse
   - posts to BB appropriate:
     - commands
     - messages
   - Waits for next cycle

To visualize how QP-theory works in this system, imagine a body of IF–THEN rules:

IF A THEN B,
IF X THEN Y,
IF Z THEN M, etc.

The parser's job is to find the IF-side (antecedent field) that best matches the current data input conditions. This may be thought of as the "recognize" part of a recognize–act cycle.

The thinker's job is to verify the parse, then set up the THEN-side (consequent field) to fire. This may be thought of as the "act" part of a recognize–act cycle. Part of the consequent field will result in changes to information posted on the blackboard, e.g., current state may have changed. Other parts of the consequent field may request control commands to be issued to the external environment.

The blackboard monitor's jobs include executing the control commands that result from firing the THEN-side, and maintenance of the flow of data to the parser and the thinker.

4.3 QP-Theory Knowledge Bases

In a QP-theory knowledge base:

The antecedent field is the same as the *preconditions* for a given process-instance.
The consequent field is the same as the *effects* for a given process-instance.

Thus, a process-instance can be thought of as an IF–THEN definition, although it is not written in that form. To visualize how a complete *user knowledge base* is constructed, the following listing illustrates the order in which components of such a program are ordered in the source code:

User Program Source Listing Diagram

```
* * * * * * * * * * * * * * * * * * * * * * *
      (direct the following to blackboard monitor)
      messages
      knowledge base
      etc.
      .
      .
      (direct the following to parser and to thinker)
      knowledge base
      etc.
      .
      .
      .
(end of user knowledge base)
* * * * * * * * * * * * * * * * * * * * * * *
```

A typical user knowledge base will have one knowledge base (KB) written for the parser and one for the thinker. Goals posted on the BB will direct the knowledge bases to behave differently.

A user knowledge base is written with a text editor then compiled on top of the sensor fusion shell. The sensor fusion shell is, itself, a knowledge base of precompiled primitive notions (e.g., behaviors, methods, and object definitions), along with a library of predefined processes (e.g., evaporation, newtonian motion, chemical reactions).

Figure 7 summarizes the contents of the sensor fusion shell. This shell also includes the blackboard monitor, and all necessary computer operation routines such as communications, sensor interface operations, and interface to a computer operator. The shell is "booted" directly into the computer hardware before compiling the user knowledge base.

A user's knowledge base, as illustrated in Figure 8, also includes extensions to the components already discussed that involve definitions (e.g., episode and history) required to describe the experiment and its various processes.

The sensor fusion system is implemented in a fully extensible computer language called FORTH. Any program elements not included in the shell, but deemed important to the experiment being conducted, may be directly added to the system. Thus, the user's

BEHAVIORS
METHODS (IF ...THEN RULES)
ACTOR DEFINITIONS

PROCESS LIBRARY
- MOTION
- HEAT
- CHEMICAL REACTIONS
- ETC

FIGURE 7 Sensor fusion shell.

EXTENSIONS TO:
- BEHAVIORS
- METHODS
- DEFINITIONS
- PROCESS LIBRARY

PROCESS-INSTANCES
EPISODES
HISTORY
GOALS

FIGURE 8 User's knowledge base.

knowledge base is, in fact, just an extension of the sensor fusion shell.

4.4 Sensor Fusion Program Structure

It is useful to consider the hierarchical structure designed into the architecture of a sensor fusion system. This consideration may be viewed both from the *bottom-up*—lowest knowledge level first—and the *top-down*—highest knowledge level first.

From the top-down perspective, every *process* in a sensor fusion system is conducted by *actors*. Every actor has a script of *methods*, which, when used (fired), describe the *behaviors* a given actor is capable of. Thus, a knowledge base in a sensor fusion system is layered top-down from the process and actors in the experiment to their behaviors. We now describe these layers of knowledge in a sensor fusion system from the bottom-up perspective. Each of the following sections describes an element of either the sensor fusion shell (Figure 7) or the user's knowledge base (Figure 8), or both.

4.1.1 Primitive Behaviors

Figure 9 illustrates the notion of a primitive behavior. Such a behavior may be thought of as a "muscle" in the system. Whenever some actor in an active process demands some calculation, or some

PERFORMANCE ROUTINES:

- HARDWARE DEVICE DRIVERS
- NUMERIC ALGORITHMS
- OTHER HARDWARE / SOFTWARE PROCEDURES

```
EXAMPLE BEHAVIOR:

HEATER-ON
PROCEDURE: HEATER-ON
     BEGIN
               OUT (HEAT-UP)
     END
:HEATER-ON (HEAT-UP) MESSAGE SEND;
```

FIGURE 9 Primitive behaviors.

communication, an appropriate primitive behavior is "fired" or activated. Such behaviors are, in fact, procedural performance routines. They include hardware device drivers (e.g., communication with the sensor interface board, serial port drivers, etc.), number crunching routines (e.g., drivers for a numeric coprocessor; whole routines for, say, Fourier transforms; etc.), or any other hardware drivers or software procedures capable of being considered as muscles of the system. An example of a primitive behavior would be a routine to turn on (or off) an autoclave heater.

4.4.2 Methods

Methods are conditional relations between the actors and the primitive behaviors available to them. As an example, in Figure 10, the behavior "turn the heater on" is available to the "temperature" actor on the condition that this particular actor has as a current goal to increase a sensed temperature value.

Methods are implemented as IF–THEN rules that read:

IF a goal is to *do something*
THEN use this *behavioral procedure*

BIND THE NEEDS OF ACTORS TO BEHAVIORS:

- **HARDWARE DEVICE DRIVERS**
- **NUMERIC ALGORITHMS**
- **OTHER COMPUTATIONAL HARDWARE / SOFTWARE**

EXAMPLE METHOD:
HEAT-UP IF GOAL IS TO RAISE TEMP NR. 1 THEN TURN HEATER-ON

FIGURE 10 Methods.

4.4.3 Primitive Definitions—Actors

Figure 11 illustrates the notion of primitive actor definitions. Using a “frame” structure, a class of actors is defined. An example of a class of actors could include the group of temperature sensors. Such sensors represent the entire concept of *temperature*, so it is reasonable to name this class “temperature.”

Instances of the class “temperature” would include each of the individual temperature sensors located in the experiment. Each individual temperature sensor represents the sensed result of methods and behaviors used to achieve changes in the experiment.

It is instructive to consider the “gestalt” notion of an actor here. We use the temperature class of actors as an example. In a given experiment, it may be desirable to increase the temperature of some part of the experiment. To do so, one first considers the nature of the notion of increased temperature: additional heat energy is required, and a change will be noticed in the sensed data. Combining these notions, notice that by defining a class of actors for the temperature concept, we are able to encapsulate both the sensed data and the methods necessary to achieve goals for that class.

DEFINES CLASSES OF ACTORS IN A PROCESS

DEFINITION INCLUDES:

- CHARACTERISTICS
- EXAMPLES / INSTANCES-OF
- METHODS

EXAMPLE DEFINITION:
TEMPERATURE (A CLASS OF SENSORS) IS-A: QUANTITY INSTANCES: (.. TEMP NR. 1 ..) METHODS (.. HEAT-UP ..)

FIGURE 11 Primitive actor definitions.

Thus, if it is desired to increase the temperature sensed at location, say, #1 in an experiment, we define an *instance* of temperature: Temp #1. This instance is, itself, an actor that has methods for taking its own reading (e.g., finding its own value in the current data input stream from the sensor interface board), and changing its own value (e.g., turning on the heater to add the additional heat energy required to satisfy some higher goal). The gestalt of this approach to system design is that, with one simple notion of an actor, we are able to capture all the classical procedures necessary to deal with any portion of an experiment as simple declarations of methods for instances of actors.

4.4.4 Process Library

Actors are the players in an experiment. They have methods for dealing with any desired goal. Their goals come from "higher" entities in the hierarchy: process descriptions.

Figure 12 illustrates the notion of a process library. A process library is a collection of process descriptions. An example of a

BINDS PROCESS-INSTANCES
TO
PRIMITIVE BEHAVIORS

THROUGH ACTORS AND THEIR METHODS

EXAMPLE PROCESS:
CHEMICAL REACTION PRECONDITIONS: (. . . HIGH TEMPERATURE . . .) EFFECTS: (. . . REACTION . . .)

FIGURE 12 Process library.

process description is that of, say, a chemical reaction. This is a generic process description of the type of chemical reaction that occurs at elevated temperatures (in general, it is prevented by freezing the chemical constituents, and is encouraged by elevating the temperature of the chemical constituents).

Such a chemical reaction requires—as a *precondition*—an elevated temperature, and will result—as an *effect* of the reaction—in a chemical reaction detectable in an "exotherm" and a change in the viscosity of the chemical constituents. Remember, this is a "generic" definition of a chemical reaction: it will shortly (in the next section) be specialized by declaring an instance (epoxy-cure) in which specific actors (e.g., temperature sensor #1) are declared.

As a generic process definition, it is part of the sensor fusion shell process library. This definition provides a coupling between a class of processes (chemical reactions) and a class of actors (temperature). Thus, it provides the coupling necessary to achieve goals in an experiment (e.g., to have a chemical reaction) and the methods and behaviors necessary to achieve those goals (e.g., turn on the heater to start the chemical reaction).

4.4.5 Process Instances

Figure 13 illustrates the notion of a process instance. An instance of an entity is a specialization of that entity. Thus, a process instance is a special declaration—a particular definition of a process that is a member of a class of processes.

As an example, the process "epoxy-cure" is an instance of the class of "chemical reaction" processes. By declaring epoxy-cure to be an instance of a class of processes, many, if not all, of the necessary behaviors will already be described in the process library entry. Thus, the process instance—in most cases—needs only to declare specific actors (e.g., temperature sensor #1), specific preconditions, and specific effects as desired.

4.4.6 Episodes

Throughout the duration of a given experiment, there may be one or many different processes involved. Some processes may occur at different times along the "history" of an experiment, and at some times, several processes may be active together. Units of time along the history of an experiment are called *episodes*, as illustrated in Figure 14.

AN INSTANCE OF A PRE-DEFINED PROCESS

LINKS A GOAL TO NECESSARY BEHAVIORS

EXAMPLE PROCESS INSTANCE:
EPOXY-CURE PRECONDITIONS: (. . . HIGH TEMPERATURE NR. 1 . . .) EFFECTS: (. . . CURE . . .) QUANTITIES: (. . . TEMP NR. 1 . . .)

FIGURE 13 Process instance.

A COLLECTION OF

PROCESS INSTANCES

EXAMPLE EPISODE:

CURE
(. . . EPOXY-CURE . . .)

FIGURE 14 Episode.

As an example of an episode, the process "epoxy-cure" may constitute the only active process. This active process may involve the temperature and viscosity actors, along with their methods and behaviors. The epoxy-cure process is comprised of the episode named "cure."

An episode is implemented as a list. On that list are all of the processes active during the episode. A list may have one or more entries, and the entries may or may not be ordered. Ordering in an episode does not deal with the "time sequence," since all processes in an episode are active at one time or another during the time period of the episode. Ordering does permit listing of processes in order of their "importance"; by listing the processes with critical actors first (e.g., pressure-related actors that may burst a pressure vessel), one causes the sensor fusion system to "consider" or "think about" those processes first.

Other episodes, coupled with the "cure" episode, will be seen (below) to constitute the *history* of the experiment.

4.4.7 History

A history is a collection of episodes. This collection, illustrated in Figure 15, is an ordered list. The ordering in a history is temporal in nature; the first element of a history list is the first episode expected to become active.

AN ORDERED SET OF EPISODES DESCRIBES ENTIRE PROCESS USED BY PARSER

EXAMPLE HISTORY:
LAMINATE-CURE (PRE-CURE CURE POST-CURE)

FIGURE 15 History.

As an example, a "laminate-cure" history includes the episodes "precure," "cure," and "postcure." Each of these episodes must become active in the proper time sequence in order for the laminate cure experiment to have a chance at success.

The history list is a principle part of the parsing process. The ordering of entries in a history list provides information about what should be active "now" and what to expect "next." By asking for a history list of what to expect next, the parser is able to correlate changes in the input data values with the onset of new episodes. The "next" entry of a history list forms an *expectation* of the next active episode, which in turn forms a definition of the expected next process—instances to become active, each of which defines its own preconditions as elements of the input data stream. Thus, by monitoring the history list, the Parser forms continuous expectations on the input data stream, looking for active processes.

4.4.8 Goals

When designing an experiment, the designer lists the goals of that experiment. Such goals should include:

What is to be achieved
What is to be avoided
What to do in case of an error

Figure 16 illustrates goals as a list ordered in the same sequence just listed. This list includes goals to achieve as a sublist of

A LIST OF:

- WHAT TO ACHIEVE
- WHAT TO PREVENT
- WHAT TO DO IN CASE OF ERRORS

EXAMPLE GOALS:
LAMINATE-CURE-GOALS (ACHIEVE (LAMINATE-CURE) PREVENT (VOIDS EXOTHERM) ON-ERROR (. . . .))

FIGURE 16 Goals.

processes, goals to prevent as a sublist of processes, and a list of error-handling processes or procedures.

Ordering of the goal list may be arbitrary, or may be according to criticality. If a type of error condition in an experiment creates dangerous conditions (e.g., overpressure bursting a pressure vessel), then the on-error list may be ordered first on the goal list to gain access to those behaviors first. In any event, ordering does not imply a temporal sequence: all goals are active at once.

The goal list is used by the thinker system. This system is charged with bringing about the behaviors necessary to achieve the goals; it uses the goal list as its "agenda." It considers all elements of the list as active: it deals with achievement goals, with prevention goals, and is always ready to make the error list active if some error is detected.

4.5 Developing a Knowledge Base

4.5.1 Introduction

The following sections presume certain prior knowledge. Documents needed in support of this knowledge include:

FORTH 83 language programming manuals
Expert 5 system manuals
Hardware manuals as required

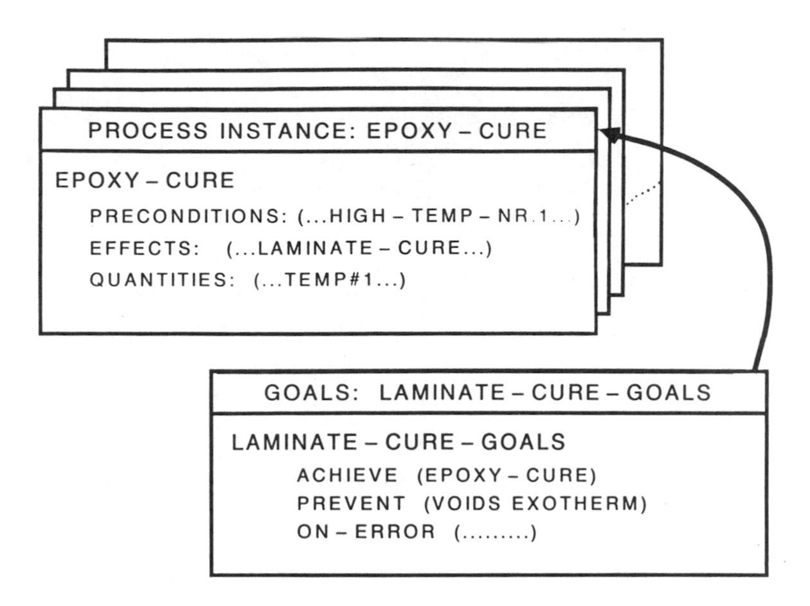

FIGURE 17 Thinking: goal to process coupling.

Any user knowledge base written for sensor fusion is effectively a coupling between desired goals and the primitive behaviors necessary to achieve those goals. Figure 17 illustrates the link created between a process-instance (epoxy-cure) and an achievement goal (epoxy-cure). The sensor fusion system "notices" that an entry on the achievement goal list is epoxy-cure, and sets the thinker program to seeing that the epoxy-cure process instance becomes active and successful.

Designing a sensor fusion knowledge base is an instance of a standard factorial design problem. In such a factorial design problem, one takes all the independent variables—in this case, all the goals—and designs an experiment that has enough instances of processes to satisfy all those goals, enough classes of actors to satisfy all those process instances, and enough methods and behaviors to satisfy all those actors.

4.5.2 Overall Program Design

The overall program design process involves:

Listing all the goals
Listing all the process instances to achieve goals

Listing all the actors in the processes
Listing all the methods used by the actors
Listing all the behaviors required for methods
Designing the blackboard entries

This list of design activities illustrates factorial design in a nutshell. If one simply starts at the top (this is top-down programming), listing all the goals—one each on a separate sheet of paper—and listing all the process instances for each goal on that goal's sheet, and, on a separate sheet of paper, all the actors, methods, and behaviors, one achieves a powerful listing of all the requirements for a program to conduct an experiment. If this process is done on a large sheet of paper, a bushy tree may appear when all the goals are linked downward to their respective processes, actors, and methods.

Figure 17 illustrates the beginning of this process, except in reverse order: goals are listed last. This is because a program for Sensor Fusion is actually written "bottom-up." To clarify this point, notice that the program is *designed* top-down: starting with the goals, working downward toward primitive behaviors. But the program, once designed, is *written* bottom-up: starting with primitive behaviors, working upward toward goals.

To summarize the notions involved in the development of a user knowledge base for the sensor fusion system, one builds a computer program that links the desired goals of the experiment with the computer behaviors necessary to achieve those goals.

4.5.3 Programming the Thinker

Figure 17 starts the Thinker design process. Starting with the list of goals, each process instance is then defined. The example (Figure 17) process-instance is epoxy-cure.

Each of the preconditions, effects, and quantities for that process instance (and all other process instances of a given goal) is listed in frames as illustrated. A typical precondition is chem-react, short for chemical reaction.

Chemical reaction is a member of the process library included with the sensor fusion system. If it were not a member of that library, it could be added to the library as one of the extensions a typical user might add.

Figure 18 takes the discussion of the thinker program from the

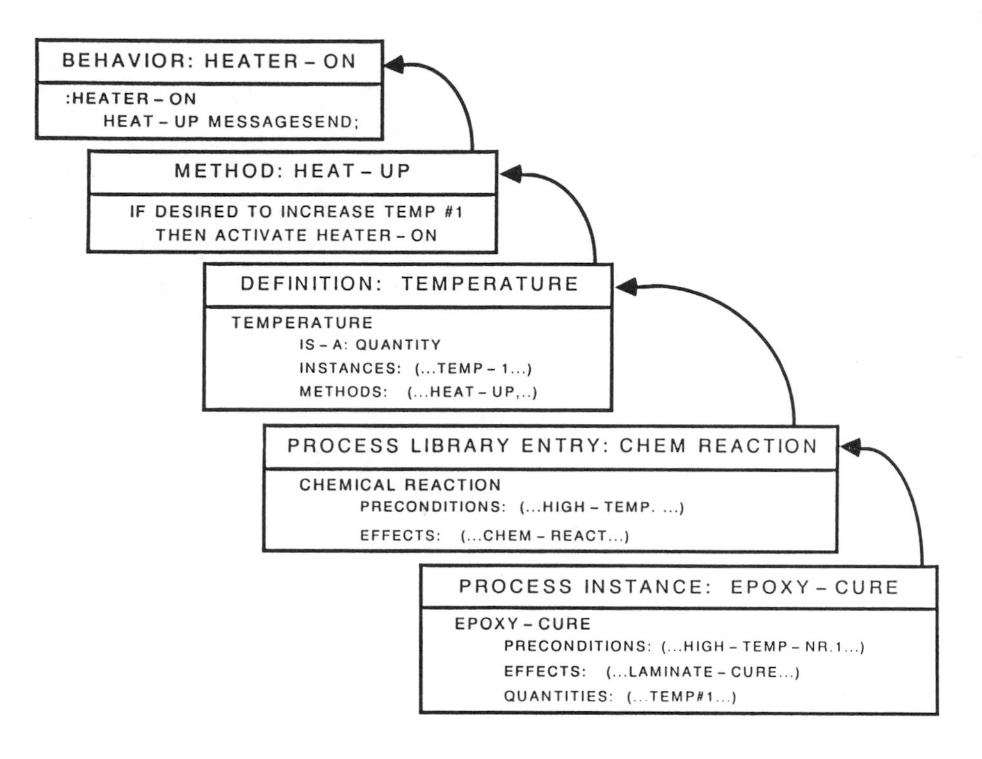

FIGURE 18 Thinking: process to behavior coupling.

process instance all the way back to a typical behavior. Following this diagram, it is possible to notice the mechanism by which the sensor fusion system becomes a network of interconnected goals, processes, actors, methods, and behaviors. A precondition for epoxy-cure is a chemical reaction. A precondition for this type of chemical reaction is elevated temperature. And temperature is an actor with a method for achieving an elevated value and a computerized behavior to satisfy that method.

Any other precondition listed with any process would require other actors—or, for that matter, the same actors with different subgoals, methods, and behaviors.

To recap the design of the thinker program, start with goals, list the process instances required to meet those goals, list the processes necessary to fulfill the preconditions of the process instances, list the actors involved in the preconditions of the active processes, and then list the methods and behaviors needed by the actors.

4.5.4 Programming the Parser

While the thinker is tasked to fulfill a list of goals of an experiment, the parser is tasked to track the history of the experiment. Like the thinker program, the parser is also designed top-down. However, instead of a list of specific achievement and prevention goals, the parser starts with an expected history of the experiment.

Figure 19 illustrates a linkage built from a history of a laminate cure cycle (the experiment) back to a process instance. Notice that the process-instance is the very same "epoxy-cure" process-instance found in the thinker program. Tracking backward from the temporal aspects of the experiment (the history, and the episodes) to the mechanical aspects of the experiment (the processes, actors, and behaviors), one can observe that the program for the parser is the same program used by the thinker, with the lone exception of the top-level declarations: the thinker couples its knowledge to a set of

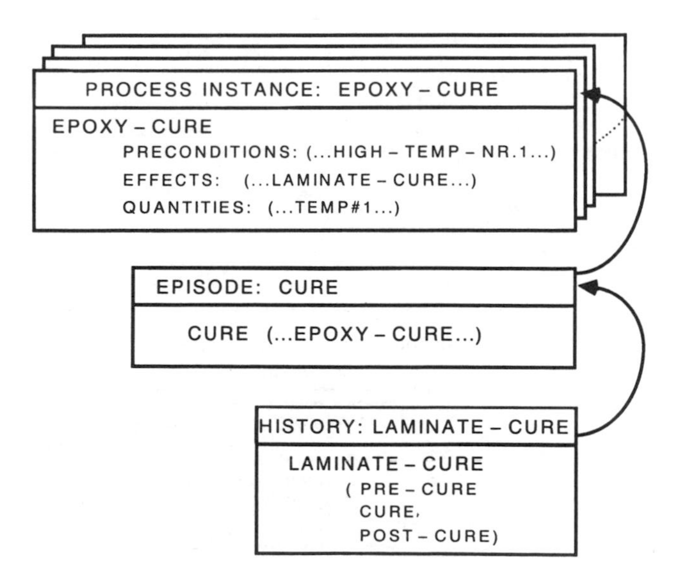

FIGURE 19 Parsing: history to process coupling.

goals, while the parser couples fundamentally the same knowledge to an expected history of the experiment.

From Figure 19, we see that the experiment "laminate-cure" includes an episode of "cure." This episode includes an active process instance "epoxy-cure." In Figure 20, we trace this lineage further back. "Epoxy-cure" has as its precondition a chemical reaction, which requires an elevated temperature. Once again, the actor "temperature" is active; this time, however, its behavior is not one of achieving an elevated temperature, but rather one of determining whether the temperature value is elevating. Different method and behavior; same actor. Parsing is the determination of the meaning of an input data stream. Thus, from this example, the parser is seen trying to answer the question "is epoxy-cure active?" by backward-chaining from the question to the primitive behaviors

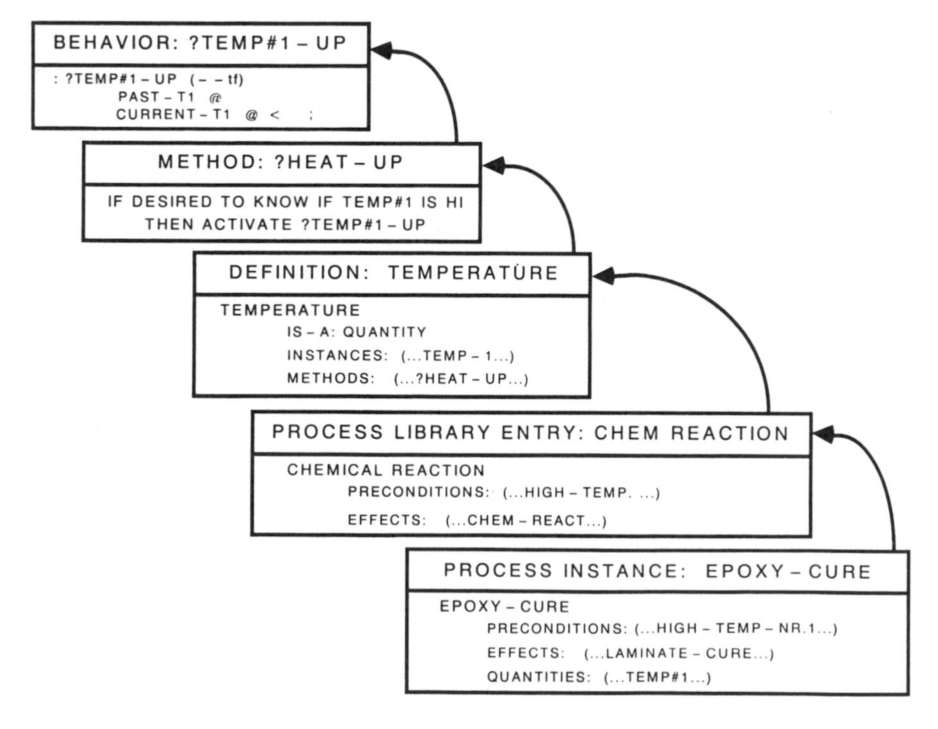

FIGURE 20 Parsing: process to behavior coupling.

that provide answers. Backward-chaining is an inference process available in Expert 5 that starts with a goal question to answer and searches backward toward the most primitive notion in the knowledge base that can support the desired answer. Each actor has methods capable of both causing change in the process, and monitoring that change. Each actor is responsible for its own methods and behaviors.

To recap the development of the parser program, the same set of declarations built for the thinker program is used for the parser, along with additional methods and behaviors needed to satisfy the special requirements of the parser.

REFERENCES

Bobrow, D. G., et al., (1985). *Qualitative Reasoning about Physical Systems.* MIT Press, Cambridge, Mass.

Bullers, W. I., S. Y. Nof, and A. B. Whinston (1980). Artificial intelligence in manufacturing and control. *AIIE Transactions, 12*(4), 351–363.

Garrett, P., Lee, C. W., and LeClair, S. R., (1987). *Qualitative Process Automation vs. Quantitative Process Control*, American Control Conference Proceedings, June 1987, Minneapolis, Minn.

LeClair, S., Lagnese, T., Abrams, F., Lee, C., Park, J., (1987). *Qualitative Process Automation for Autoclave Curing of Composites*, AFWAL-TR-87-4083, May 15, 1987, Wright-Patterson AFB, Ohio.

Ni, H. P. (1986). Blackboard systems: The blackboard model of problem solveng and the evolution of blackboard architectures. *The AI Magazine*, Summer, 38–53.

Park, J. (1985). *Expert-5 Programmer's Guide*; ThinkAlong Software Inc, Brownsville, Calif.

SECTION VI

Real-Time Machine Tool Control

13
A Real-Time Control System for a CNC Machine Tool Based on Deterministic Metrology

M. Alkan Donmez*
The Catholic University of America
Washington, D.C.

This paper describes a real-time control system for a Computer Numerical Control (CNC) machine tool. This system is used to compensate for the machine tool geometric and thermally-induced errors in real-time during the cutting operation. The errors are predicted based on a combination of kinematic and empirical models. Empirical models are obtained by carrying out series of

*Current affiliation: National Bureau of Standards, Gaithersburg, Md.

Parts of this article have already appeared in the following publications: *Proceedings of the 1986 IEEE International Conference on Robotics and Automation; Modelling, Sensing and Control of Manufacturing Processes,* PED–vol. 23, 1986, Production Engineering Division of ASME; and *Precision Engineering,* January 1987. Based on a presentation made at "Statistical Process Control: Keeping Pace with Automated Manufacturing, a National Symposium," sponsored by the Center for Professional Development and the Reliability, Availability and Serviceability Laboratory, College of Engineering and Applied Sciences, Arizona State University, November 6–7, 1986.

measurements on the machine tool. This system is implemented on a two-axis turning center, and significant accuracy enhancement is achieved on the machined parts.

1. INTRODUCTION

In today's computer-aided manufacturing environment, one of the main objectives of quality control and assurance programs is to achieve higher accuracy of the machined workpieces. While there are various sources of errors that result in unacceptable parts, eventually the accuracy of the workpiece, in metal-cutting operations, depends on the accuracy of the relative position between the cutting tool and the workpiece (Hocken et al., 1977; Schultschik, 1977; Dimensional Metrology Group, National Bureau of Standards, 1979; Love and Scarr, 1973). Therefore, one way to improve the workpiece accuracy is to utilize concept of deterministic metrology by focusing on the machine tool and the cutting process to control the cutting tool position accurately in the workspace of the machine tool. This paper describes the application of deterministic metrology to a Computer Numerical Control (CNC) machine tool to develop a real-time control system for compensation of machine tool geometric and thermally induced errors. This control system is used to improve the machine tool accuracy for precision machining.

Commercially available CNC machine tool controllers use closed-loop position servo systems to locate the cutting tool with respect to the workpiece in the machine tool workspace. In general, encoders or resolvers are used as position feedback devices in these controllers. These position feedback devices are usually attached to the slide driving mechanisms such as the ballscrew/nut assemblies. Since there is an offset between the axis of measurement and the cutting tool (which is called Abbe offset), any rotation and/or translation between the feedback device and the cutting tool is not detected by the servo system of the machine tool. Furthermore, since most machine tools have stacked slides, any small error in one slide motion is amplified by the motion of the adjacent slide.

The factors that cause these small rotations and translations are classified (Hocken, 1980; Weck, 1980; Spur and DeHaas, 1973; Sato, 1974; Tlusty and Mutch, 1973; McClure, 1980; Sata et al., 1973) as

1. Geometric errors
2. Thermal effects
3. Static loading

Geometric errors are caused by the imperfections created during the manufacturing of the machine tool elements and the misalignments created during the assembly stage of the machine tool. These types of errors are systematic in nature and can be compensated for if determined accurately. Similarly, the end effect of thermal deformations in the machine structure, which are due to the temperature gradients created by the heat generated during the cutting operation, are small rotations and translations that can be determined systematically. Finally, static loading, which is created by the cutting forccs, also causes deformations in the workpiece and cutting tool as well as the machine structure. Although it can be found deterministically, the effect of static loading is beyond the scope of this chapter. The purpose of the real-time control system described here is to interact with the CNC controller of the machine tool and compensate for the machine tool geometric and thermally induced errors. The mathematical and the empirical error models used for predicting the error of the cutting tool position, as well as the error compensation system itself, are described in the following sections.

2. GENERATION OF ERROR MODELS FOR REAL-TIME COMPENSATION

As a result of the above-mentioned error sources, there is always a difference between the actual tool position and the desired (ideal) tool position. This difference in position can be represented by a vector in the machine tool workspace. A rigid-body kinematic analysis is used to determine the positional error vector of the cutting tool. In this analysis, the machine tool–fixture–workpiece system is considered as a chain of linkages with spatial relationships between them. These spatial relationships include error motions of each linkage. There are six types of error motions: three translational errors along each orthogonal axis (linear displacement and straightness in two orthogonal axes) and three rotational errors about these axes (roll, pitch, and yaw). These error motions are shown in Figure 1. A homogeneous coordinate transformation

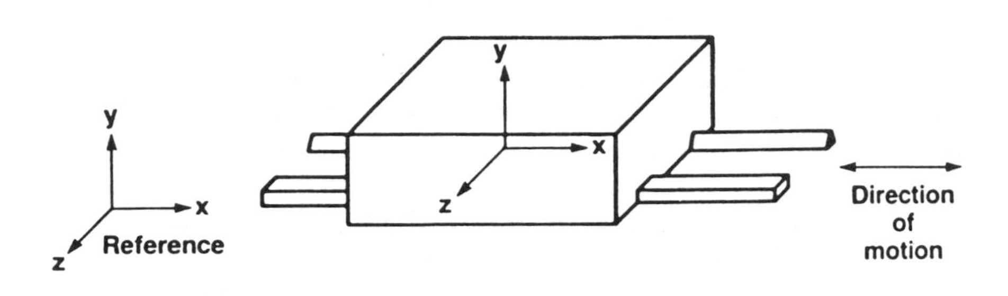

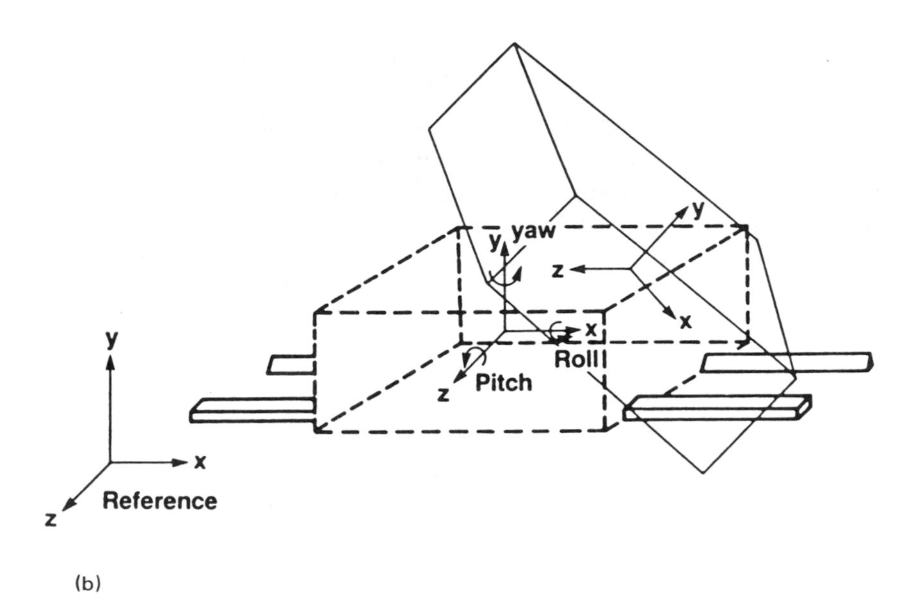

FIGURE 1 Six degrees of freedom error motion of a typical machine slide: (a) ideal slide and (b) actual slide.

matrix describes the position and orientation of a rigid body with respect to another. Such a transformation matrix assigned to each linkage represents all six types of error motions in the following form:

$$T = \begin{bmatrix} 1 & -\varepsilon_z & \varepsilon_y & \delta_x \\ \varepsilon_z & 1 & -\varepsilon_x & \delta_y \\ -\varepsilon_y & \varepsilon_x & 1 & \delta_z \\ 0 & 0 & 0 & 1 \end{bmatrix} \tag{1}$$

where ε terms are rotational and δ terms are translational errors.

By assigning a homogeneous coordinate transformation matrix to each element of the machine tool–fixture–workpiece system, it is possible to construct a homogeneous transformation matrix equation corresponding to the structural loop of the system. An example of such an equation is

$$[T_s][T_w] = [T_z][T_x][T_t][E] \tag{2}$$

where $[T_s]$ is the homogeneous coordinate transformation matrix for the spindle, $[T_w]$ is the homogeneous coordinate transformation matrix for the workpiece, $[T_z]$ is the homogeneous coordinate transformation matrix for the carriage, $[T_x]$ is the homogeneous coordinate transformation matrix for the cross slide, $[T_t]$ is the homogeneous coordinate transformation matrix for the tool turret, and $[E]$ is the resultant error matrix.

In Eq. (2), the resultant error matrix E includes the positional error vector consisting of individual error terms of each contributing element of the system. The following equations show the components of the resultant positional error vector p_E when Eq. (2) is solved for the error matrix E.

$$\begin{aligned} p_{Ex} = {} & x(w) + \delta_x(w) + \varepsilon_y(s)z(w) + \delta_x(s) - X_4 - \delta_x(c) \\ & - [\varepsilon_y(z) + \varepsilon_y(x) + \varepsilon_y(t)]Z_4 \\ & - \delta_x(t) - x - X_1 - \delta_x(x) - \delta_x(z) \end{aligned} \tag{3}$$

$$\begin{aligned} p_{Ey} = {} & \delta_y(w) - \varepsilon_x(s)z(w) + \delta_y(s) - [\varepsilon_z(z) + \varepsilon_z(x) + \varepsilon_z(t)]X_4 \\ & - \delta_y(c) + [\varepsilon_x(z) + \varepsilon_x(x) + \varepsilon_x(t)]Z_4 \\ & - \delta_y(t) - \varepsilon_z(z)x - \delta_y(x) - \delta_y(z) \end{aligned} \tag{4}$$

$$p_{Ez} = -\varepsilon_y(s)x(w) + z(w) + \delta_z(w) + \delta_z(s) + [\varepsilon_y(z) + \varepsilon_y(x) + \varepsilon_y(t)]X_4 - Z_4 - \delta_z(c) - \delta_z(t) + \varepsilon_y(z)x - \delta_z(x) - \delta_z(z) - z \quad (5)$$

In Eqs. (3), (4), and 5, $x + X_1$ and z are the nominal machine positions; $x(w)$ and $z(w)$ are obtained from workpiece geometry; $\delta_x(w)$ and $\delta_z(w)$ are functions of thermal and static load deformations; $\delta_x(s)$, $\delta_z(s)$, and $\varepsilon_y(s)$ are spindle thermal drift characteristics; $\delta_x(c)$ and $\delta_z(c)$ are functions of thermal and load deformations and wear; $\delta_x(t)$, $\delta_z(t)$, and $\varepsilon_y(t)$ are functions of angular position of the turret; and $\delta_x(x)$, $\delta_z(x)$, $\varepsilon_y(x)$, $\delta_x(z)$, $\delta_z(z)$, and $\varepsilon_y(z)$ are functions of machine tool geometry and thermal characteristics.

The individual error terms in Eqs. (3)–(5) are determined using deterministic machine tool metrology. Any error term e_i can be considered as some combination of nominal position (x) and temperature (T). This may be expressed as

$$e_i = a_0 + a_1x + a_2x^2 + \cdots + b_1T + b_2T^2 + \cdots \quad (6)$$

The order of Eq. (6) as well as the coefficients appearing in this equation are determined by measuring the errors of the machine tool and correlating them to the nominal positions and the temperatures. Since two independent variables (x and T) are considered to affect the errors, measurements are carried out over the possible ranges of these variables. For each machine axis, the measurement starts when the slide under study is at one end of its travel range. Then the slide moves toward the other end of its travel range while a reading is taken at every measuring interval. To measure the reversal error, the motion is reversed at the end of the travel, and the slide is sent back to its starting position with a reading taken at every measuring interval again. In selecting the measuring intervals, care should be taken to decouple the periodic error components due to the ballscrew misalignment. This can be done by taking the measurements at multiples of ballscrew lead. Periodic error components can be measured separately and superposed onto the main error behavior.

Due to their complex structures, machine tools, in general, require about 8–10 h of operation before they reach thermal equilibrium. Figure 2 shows the temperature profile of the machine under study. During this transient period, temperatures around the machine structure rise about 15–20° F. This rise in temperature caused errors in the order of 40 μin/°F in the radial direction and 400 μin/°F in the axial direction in the turning center under study. In order to find the thermally induced changes on the machine errors, these errors were measured over the whole temperature spectrum of the machine. The error measurements started when the machine was "cold." After five cycles of measurements, each of which consisted of movement over the whole trovel range of the axis being measured, the machine was allowed to warm up. Two types of warm-up procedure were used. In the first one, the machine was on, but there was no movement. This caused a slow warm-up. In the second one, the warm-up was induced by moving the machine slide back and forth. During each measurement cycle, in

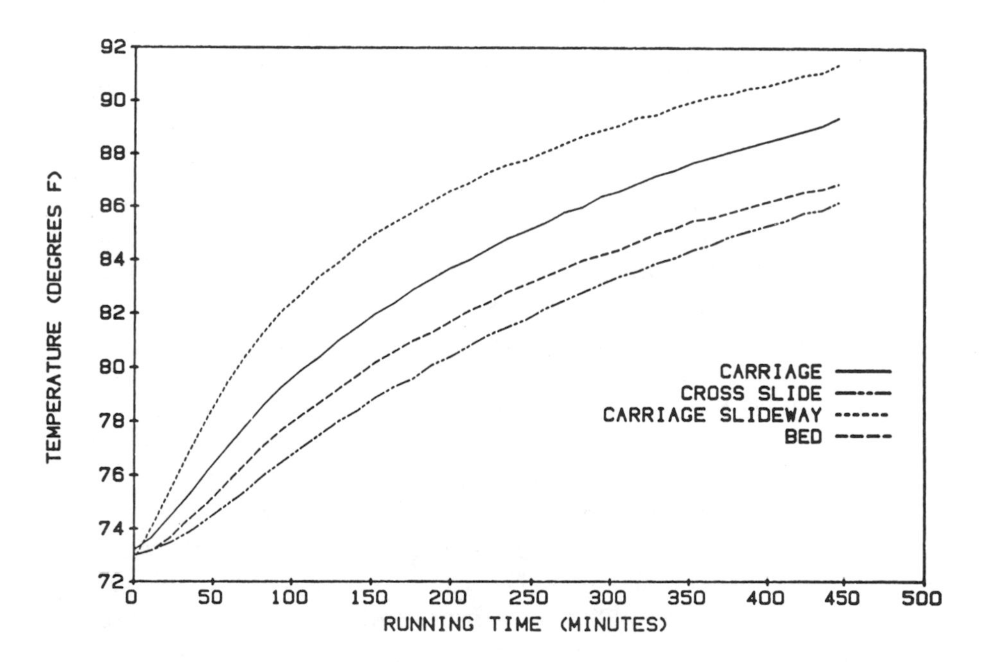

FIGURE 2 Temperature profile of the machine elements under continuous running conditions (2000 rpm, 100 ipm).

addition to the error movements of the slide under study, the temperatures of the locations around the slide were monitored. These critical locations are bearing housings at both ends of the ballscrew of the slide, the slideways, the drive motor, and the bed on which the slide is moving. During the data analysis stage, the best representative set of temperatures was chosen to be used in the calculations of individual error terms.

The error measurements are done by using a laser interferometer system and high-precision capacitive proximity sensors. The laser interferometer uses an He–Ne laser source, which emits two beams at two different linearly polarized frequencies (Estler, 1985). One of the beams is the reference beam, and the other is the measurement beam. Displacement of the moving reflector causes a Doppler shifted frequency component (ΔF) on the measurement beam, which is proportional to the amount of motion of the reflector. When two Doppler-shifted beams return to the receiver of the system, they form fringes, which are in turn counted by the receiver and converted to displacement. This type of system. with proper optics, can achieve a resolution of 0.4 microinch. The system, which was used for displacement measurements in this study, has 1 μin resolution. The resolution for the angular measurements is 0.1 arc-sec. Figure 3 shows the schematics of the displacement measurement with laser interferometry. In this study, the laser interferometer system was used for the measurement of linear displacement errors and yaw errors of the slides; the proximity sensors were used for straightness, orthogonality, and parallelism error measurements, in addition to the measurements of the spindle thermal drift in axial and radial directions. For straightness and parallelism of carriage motion, two high-precision proximity sensors were used against a precision ground test arbor, which was mounted on the spindle. However, even with a precise test arbor, one must eliminate the measurement errors caused by the nonstraight profile of the test arbor itself. This is done using reversal error technique (Donaldson, 1972). This technique requires two sets of data, taken with different sensor and artifact (test arbor) configurations, for the same type of measurement. The desired error data is obtained by algebraic summation of the two data sets. Figure 4 shows the schematic of two sets of measurements to

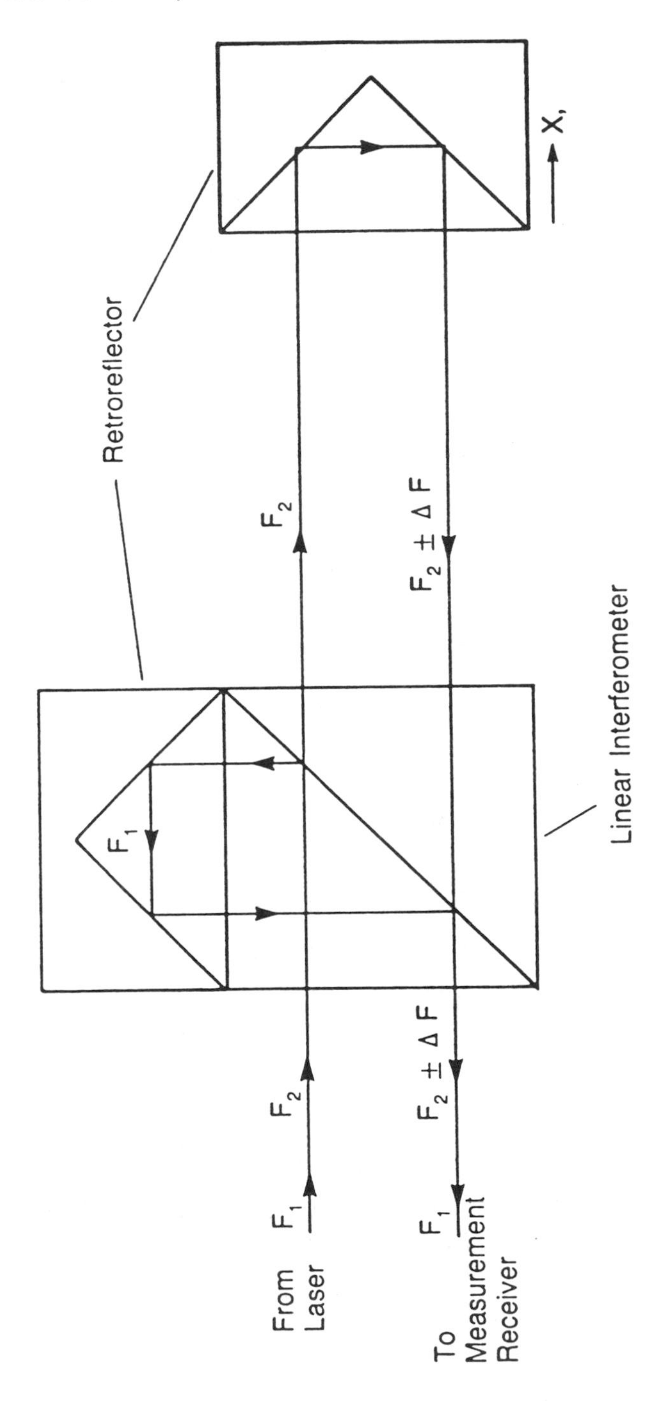

FIGURE 3 Linear interferometer and retroflector arranged for displacement measurements.

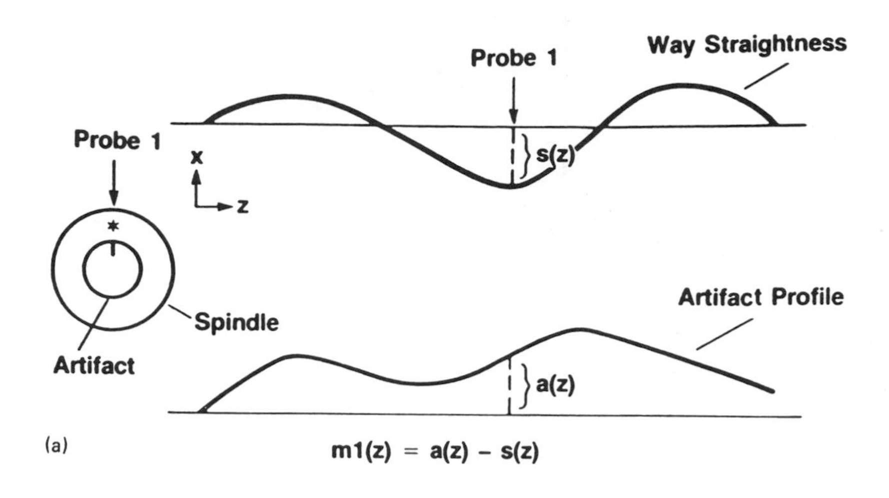

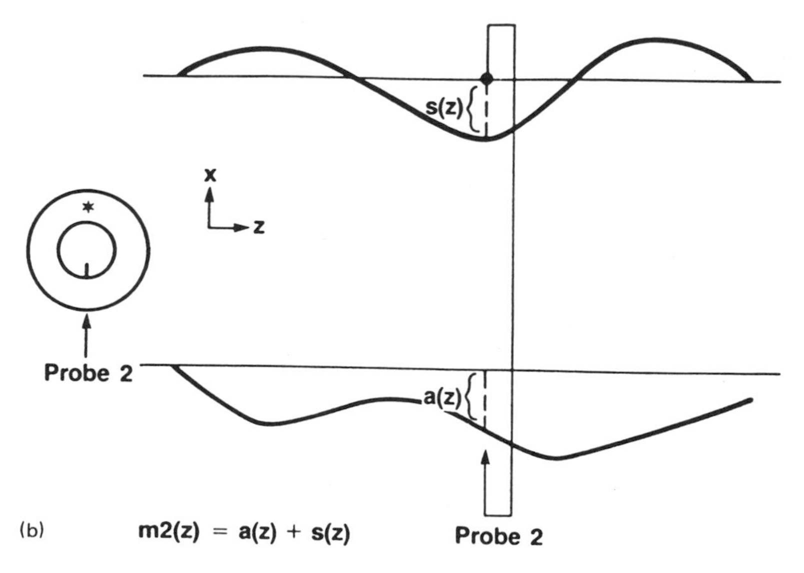

FIGURE 4 The schematic of the x straightness of the z motion measurement technique: (a) first set of measurements $m1(z)$ with probe 1; (b) second set of measurements $m2(z)$ with probe 2 (artifact rotated 180° inside the spindle).

separate artifact errors from the errors of motion. As seen in this figure, $[m2(z)-m1(z)]/2$ gives the straightness error of the carriage motion. Figure 5 shows the block diagram of the data acquisition system. Some of the measurement set ups are shown in Figures 6, 7, and 8. Figures 9, 10 and 11 show some results from these measurements and consequtive data analyses. Based on these analyses, the coefficients appearing in Eq. (6) were determined, and a set of relationships for different types of errors was generated for later use by the error compensation system.

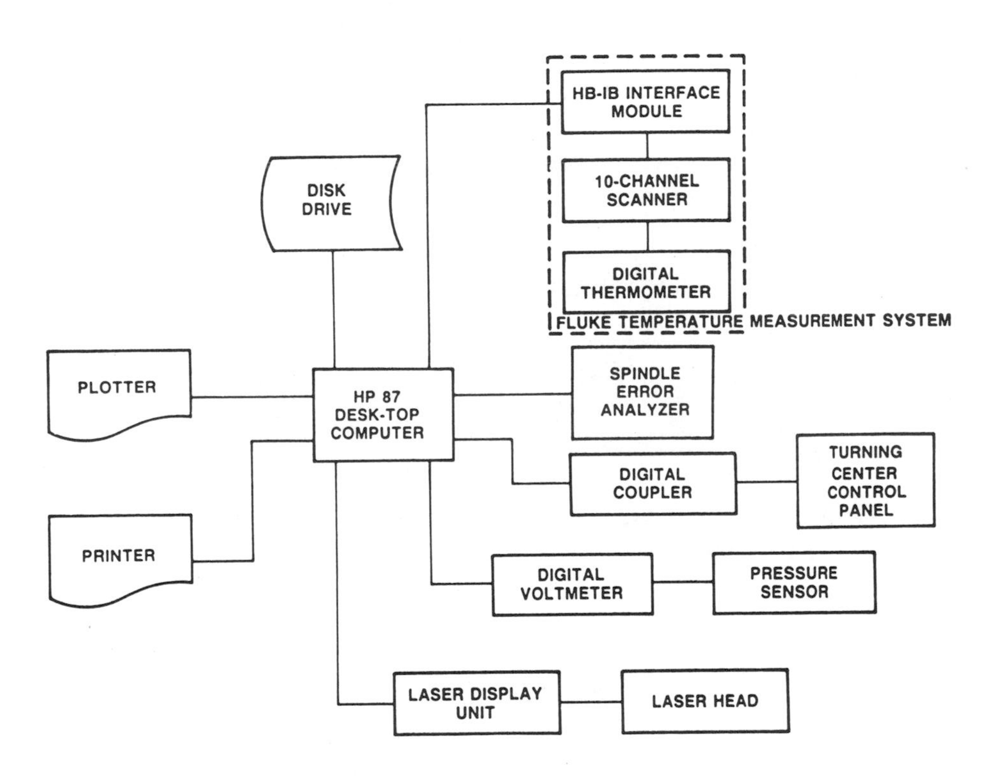

FIGURE 5 Block diagram of the data acquisition system for automatic error measurements.

FIGURE 6 The Laser interferometer set up for x displacement error measurements.

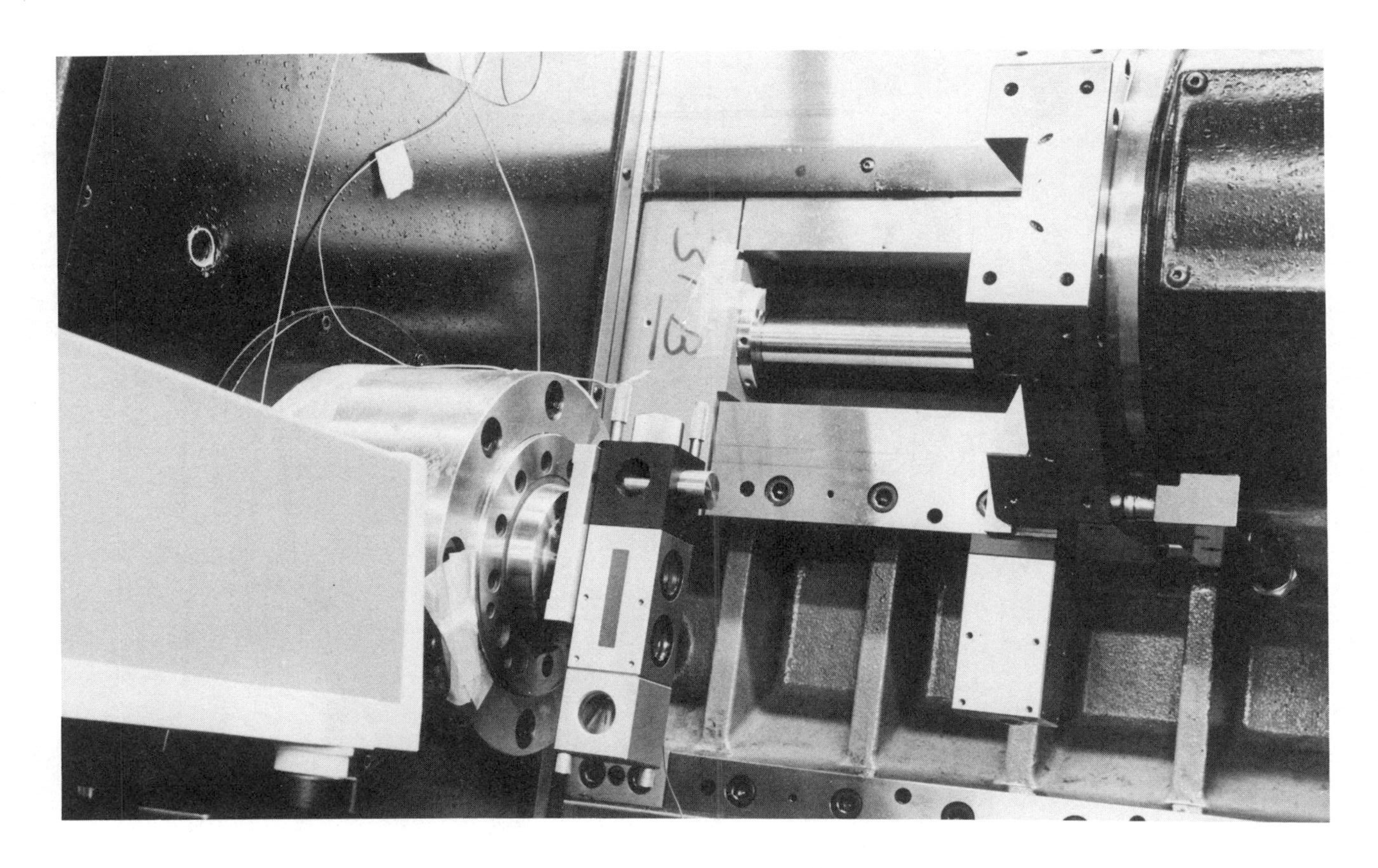

FIGURE 7 The Laser interferometer set up for z yaw error measurements.

FIGURE 8 The set up for the measurements of z straightness of x motion.

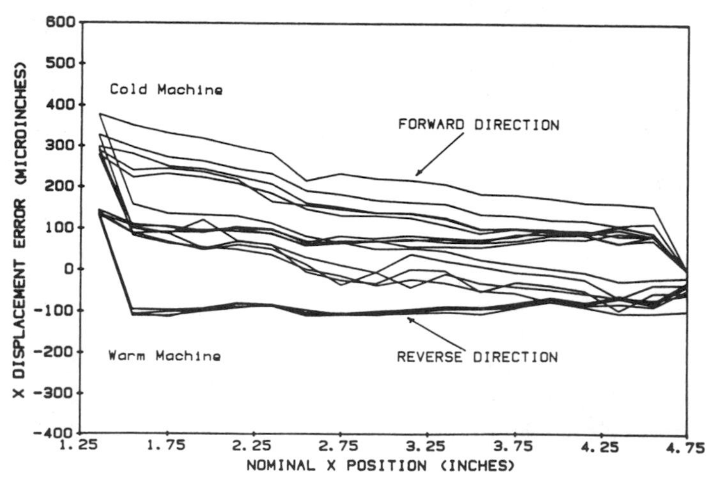

FIGURE 9 The x displacement error measurement raw data taken as the machine warms up.

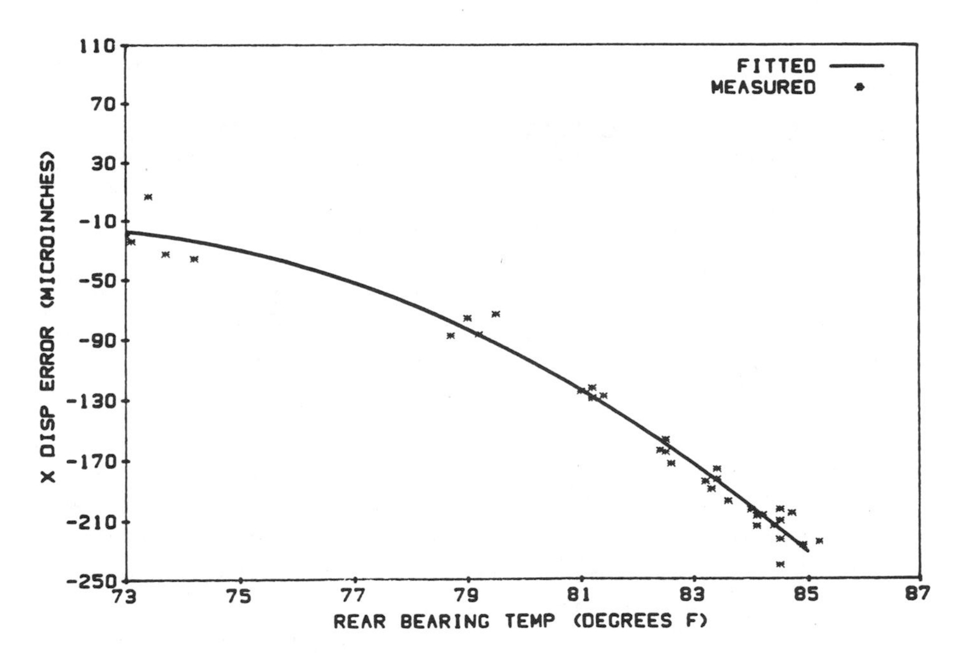

FIGURE 10 Displacement error at $x = 1.95$ in.

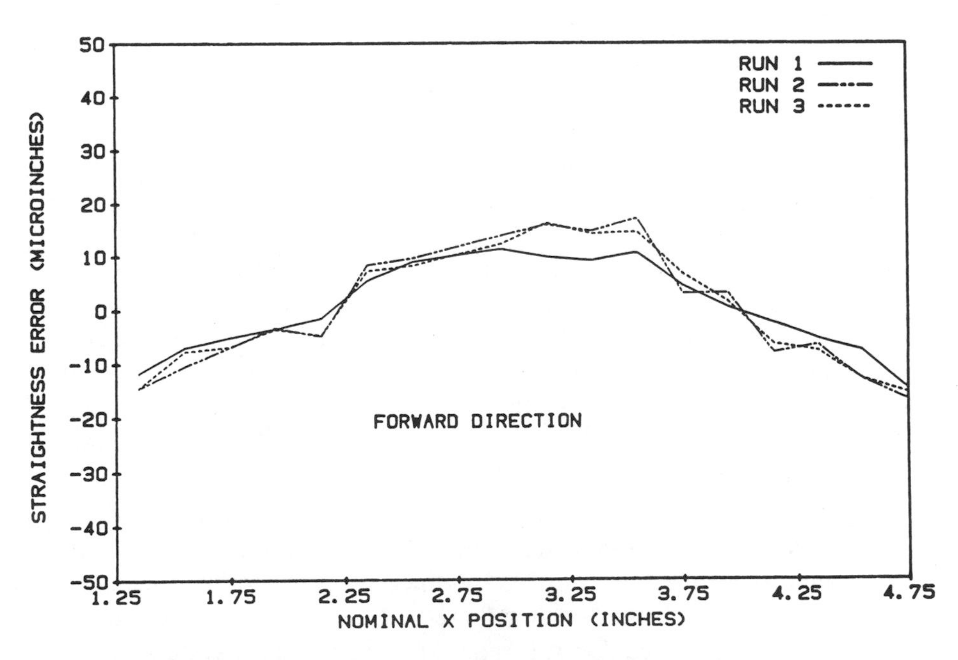

FIGURE 11 Calculated z straightness of the x motion data for the forward direction of the x motion.

3. IMPLEMENTATION OF THE CONTROL SYSTEM

The resultant error vector, which is calculated by solving Eq. (3)–(5), is used by the real-time control system to compensate for the machine tool geometric and thermally induced errors. In a typical machine tool, the axis servomotor rotates the ballscrew based on the error signal derived from the position command, the position feedback, and the velocity feedback signals. The position feedback is obtained from a position sensor such as a glass scale, an encoder, an inductosyn, or a resolver. The CNC controller calculates the position command signals, compares the command value to the position feedback signal, and performs the speed control. A real-time error compensation system is an attachment to the CNC controller of the machine tool, which injects the error compensation signals determined from the resultant error vector into a position servo loop through the registers containing the "following errors." A following error is the lag of actual position from the commanded position by an amount proportional to the velocity of the axis motion. The position command signal, summed with the following error, is the error signal used to drive the axis servo in the next servo cycle. During a cutting operation, the control system continuously receives information about the machine axis positions through a communication interface between itself and the machine's own CNC controller. Furthermore, a temperature measurement system, capable of monitoring 10 thermocouple channels simultaneously, supplies the information about temperatures of the critical locations around the machine structure, in real time. Using these two sets of data, and the error relationships previously established during the calibration stage, the control system calculates the total positional error of the cutting tool along the directions of machine axes, converts them into servo counts, and injects them into corresponding following error registers. With the error compensation system implemented, the contents of the following error registers (one for each axis) are modified at every servo cycle such that the driving signal is either increased or decreased to compensate for the calculated positional error value. The system block diagram is shown in Figure 12.

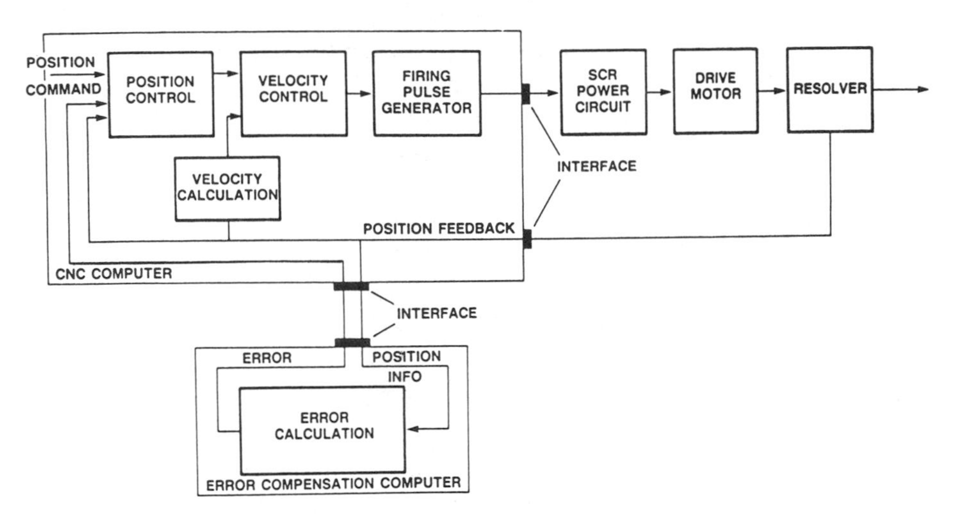

FIGURE 12 Block diagram for error-compensated CNC axis drive.

A multibus single-board microcomputer, Intel ISBC 86/30, with 128K of RAM and 64K of EPROM memory, is the main controller of the overall system. This microcomputer board contains a 16-bit 8086 microprocessor as the central processor unit (CPU), and a high-speed version 8087A numeric coprocessor for floating-point arithmetic operations. The combination of 8086 and 8087A makes it possible to run the computer at an 8-MHz clock rate to meet the requirement of high servo bandwidth necessary for contouring cuts. This microcomputer uses a multibus communications expansion board with four RS-232 serial I/O (input/output) ports to communicate with the other components of the system, such as the temperature measurement system, the tool-setting station, the keyboard interface module, and a CRT (cathode ray tube) terminal used for data entry and manual control of operations. Communications with the CNC machine tool controller are performed by three multibus parallel I/O ports, one for each axis and one for the command status protocol.

The system software is written in a high-level structured language, PLM, for flexibility and easy maintainability. It is written in a modular fashion such that the errors due to tool and workpiece deflections and other error sources that might be found in the future can be added to the resultant error calculation scheme. In order to synchronize the compensation controller with the software servo cycle of the CNC controller, a positive handshaking is established between the two controller. To do this, along with the updated error servo counts, the compensation controller sends an incremented index code to the machine controller. This index code is used by the machine tool controller to differentiate between the old and new sets of error servo counts. Since the error calculation takes about 10 msec, the new set of servo counts corresponding to a current position is received by the machine controller in the next servo cycle. Therefore, there is a maximum of two servo cycles time lag between a nominal position and the compensation execution corresponding to that position. This time lag does not cause any problem in the machining operation due to the fact that the tool motion is slow in comparison.

4. RESULTS AND CONCLUSION

In order to test the capability of the real-time error compensation system, a series of cutting test were carried out under transient thermal conditions. During these tests, under similar conditions, two sets of parts were machined, one set with error compensation applied and the other set without error compensation. The results of these tests showed significant geometric and dimensional accuracy improvements over the parts machined without error compensation. Tables 1 and 2 and Figure 10 show some results from these

TABLE 1 Error in Diameter (Nominal Diameter: 1.605 inch)

Compensated (μin)	Uncompensated (μin)	Improvement (ratio)
530	1050	1.98
270	1030	3.81
−130	1470	11.31
−150	2230	14.87

TABLE 2 Error in Length (Nominal Length: 3.44 inch)

Compensated (μin)	Uncompensated (μin)	Improvement (ratio)
−150	570	3.8
390	4410	11.31
−250	5240	20.96
450	6390	14.20

tests. The complete set of results from these cutting tests was given in Donmez (1985). While attaining higher accuracy, the system also eliminated the need for machine warm up period, which is a common and costly practice in a typical machine shop environment for precision parts.

REFERENCES

Dimensional Metrology Group, National Bureau of Standards. (1979). Annual Progress Report for Bureau of Engraving and Printing.

Donaldson, R. (1972). A simple method for separating spindle error from test ball roundness error. *Annals of CIRP*, *21*.

Donmez, M. A. (1985). A general methodology for machine tool accuracy enhancement—Theory, application and implementation. Ph.D. dissertation, Purdue University, West Lafayette, Ind., August.

Estler, W. T. (1985). High-accuracy displacement interferometry in air. *Applied Optics*, *24*(6).

Hocken, R. J. (1980). Quasistatic machine tool errors. *Technology of Machine Tools, MTTF*, *5*.

Hocken, R. J., et al. (1977). Three dimensional metrology. *Annals of CIRP*, *26*.

Love, W. J., and A. J. Scarr. (1973). The determination of the volumetric accuracy of multi axis machines. *Proceedings of 14th MTDR Conference*.

McClure, E. R. Thermally induced errors. *Technology of Machine Tools, MTTF, 5.*

Sata, T., Y. Takeuchi, and N. Okubo. (1973). Analysis of thermal deformation of machine tool structure and its application. *Proceedings of 14th MTDR Conference.*

Sato, H. (1974). Machine tool. *Bulletin of Japan Society of Precision Engineering, 8,* (2).

Schultschik, R. (1977). The components of volumetric accuracy. *Annals of CIRP, 26.*

Spur, G., and P. DeHaas. (1973). Thermal behaviour of NC machine tools. *Proceedings of 14th MTDR Conference.*

Tlusty, J., and G. F. Mutch. (1973). Testing and evaluating thermal deformations of machine tools. *Proceedings of 14th MTDR Conference.*

Weck, M. M. (1980). Geometric and kinematic errors. *Technology of Machine Tools, MTTF, 5.*

Index